SpringerBriefs in Earth Sciences

T0212750

More information about this series at http://www.springer.com/series/8897

Eugene A. Ustinov

Sensitivity Analysis
in Remote Sensing

 Springer

Eugene A. Ustinov
Jet Propulsion Laboratory
California Institute of Technology
Pasadena, CA
USA

ISSN 2191-5369   ISSN 2191-5377   (electronic)
SpringerBriefs in Earth Sciences
ISBN 978-3-319-15840-2   ISBN 978-3-319-15841-9   (eBook)
DOI 10.1007/978-3-319-15841-9

Library of Congress Control Number: 2015934044

Springer Cham Heidelberg New York Dordrecht London

Printed on acid-free paper

Springer International Publishing AG Switzerland is part of Springer Science+Business Media (www.springer.com)

*Dedicated to my parents,
Alexey and Euphalia Ustinov*

Preface

Dear reader,

If you are reading these lines, then you are holding in your hands a book on a subject that I have been pondering about for a long time.

You see, there is a sizable amount of books on remote sensing, the science, which was born before the beginning of the Space Era, and which began blossoming after the Space Era had begun.

My concern about this literature, since a long time ago, was pretty simple. In the overwhelming majority of this literature, the methods of solution of the inverse problems of remote sensing were, and still are, a predominant subject. Nobody is arguing that effective methods of computation of jacobians—matrices containing the sensitivities, i.e., derivatives of the observed parameters with respect to parameters one wants to infer—are important. But with Moore's Law still at work, and who knows for how many more years, computers become even more and more powerful, and computation of sensitivities can still be tackled simply by brute force. All you need is to buy a more powerful computer, or to add a few more processors to your cluster already in use, and run your model again and again for small finite variations of its input parameters, one at a time.

This is not the case in the area of remote sensing, which deals with measurements from orbit of radiances emanating from the atmosphere. The most desirable parameters to be retrieved are vertical profiles of atmospheric parameters, which should be specified on a grid of the vertical coordinate running through many atmospheric scale heights, resulting in (many) tens of grid values of these input parameters. That is why the methods alternative to the above brute-force approach began to be developed in the middle of the last century, and their development in this particular area of remote sensing is still going on. I am referring to the methods based on the linearization and adjoint approaches. I have also made some contribution to the development and application of the adjoint approach in this area of remote sensing.

More than a decade ago I realized that these methods—those of sensitivity analysis of models—can be generalized to be applicable to essentially any area of remote sensing, where the models (as in radiative transfer) relevant to remote sensing based on radiances measured from orbit are based on corresponding differential equations with relevant initial and/or boundary conditions.

This is not to say that the adjoint approach, for example, is not used outside the realm of radiative transfer models. There is a blossoming area of research and applications called "variational assimilation of data." A common feature, a hallmark of this approach, is the single artificial observable, called the cost function (also known under other names), which is essentially a norm of residuals—differences between measured and simulated observables. The adjoint approach is used to compute the sensitivities of this artificial observable with respect to all model parameters of interest. Since these sensitivities turn out to be proportional to the residuals themselves, the convergence of the solution of the corresponding (very nonlinear) inverse problem is notoriously slow.

Instead, I have decided to see what can be done if one keeps the individual observables, and generalizes the idea of the adjoint approach used in the radiative transfer models toward a vector of such observables. This generalization turned out to be very straightforward, and after I applied this idea to a simple zero-dimensional model of atmospheric dynamics, I moved on to sensitivity analysis of a number of models in different areas of remote sensing.

A few years ago, I realized that the well-known technique of reducing the higher-order differential equation to a system of differential equations of first order can be used as a basis of a general approach to formulation of adjoint problems with higher-order equations. The general principles of the developed approach and the results of its application to a number of stationary and nonstationary equations are presented in this book, and this is the first time they are published.

This book is intended for a variety of readers. For students of remote sensing, this book will provide an introduction to the goals, objectives, and methods of sensitivity analysis. Researchers already active in the field will find many alternative viewpoints on many particular issues of sensitivity analysis. The author understands that some statements in this book may appear arguable to particular readers, but hopefully this book will not leave the readers indifferent to its general subject, sensitivity analysis. In the author's opinion, sensitivity analysis certainly deserves more attention among researchers and certainly represents a wide research field where much more still remains to be done.

Many individuals have supported this chain of research throughout my career. I owe a great deal of gratitude to them.

It began during my tenure as a Ph.D. student at the Space Research Institute (IKI) in Moscow, Russia. Prof. Vassili Moroz was my advisor, and he guided me all the way to defending my Ph.D. thesis in 1978. Later, he invited me to join the science team of Venera 15&16/PFS instrument successfully flown in orbit around Venus.

Another productive period of my work on this research theme is associated with my tenure at the Tartu Observatory in Tõravere, Estonia. It resulted in a second Ph.D. defended at the University of Tartu in 1992. I am much indebted to Charles Willmann, Kalju Eerme, Uno Veismann, Tõnu Viik, and to all the friendly staff of Tartu Observatory for their support of my work.

In the mid-1990s, I had an opportunity to apply the results of this research to the data of remote sensing of atmospheres of Jupiter, Saturn, Uranus, and Neptune obtained from the Voyager/IRIS instrument, first at NASA Goddard Spaceflight Center and later at Cornell University. I am grateful to Barney Conrath and Peter Gierasch for this opportunity.

I continued my research in sensitivity analysis at the Jet Propulsion Laboratory (JPL), where I was invited in 1998 to participate in development of the retrieval algorithms for the TES instrument, which was later launched onboard the EOS Aura spacecraft. I am grateful to Reinhard Beer for this invitation, which opened a way for me to participate in other exciting projects at JPL while continuing my research in sensitivity analysis.

All in all, I am much indebted to all of these individuals, who, directly or indirectly, guided me during all these years. Recently, I have summarized the results of this research in the form of a short course in sensitivity analysis, which was sponsored by the Science Division of JPL and delivered at JPL. I am grateful to Geoff James for making this happen.

This particular book would not see the light, but for a kind invitation from Springer to write it. I am grateful to Petra van Steenbergen for this invitation, and to Hermine Vloemans for her kind guidance during preparation of this manuscript. I am also grateful to my JPL colleagues, Van Snyder, Phil Moynihan, Suniti Sanghavi, and Bruce Bills for reviewing the draft of this manuscript, for their suggestions regarding improving the text, and for weeding out my typos and grammar errors, inevitable for this author, for whom English is not his native language. Of course, I take responsibility for any remaining typos and errors.

And, last but by no means least, I am immensely grateful to my wife, Lyudmila, for supporting me in my endeavors in general, and during the writing of this book in particular.

Altadena, CA, USA   Eugene A. Ustinov

Contents

Chapter 1
Introduction: Remote Sensing and Sensitivity Analysis

Abstract By its basic definition, remote sensing is simply indirect measurement. In a vast variety of practical applications, we cannot measure quantities of interest directly, but we can measure some other quantities that are related to quantities of interest by some known relations. For example, targeted measurements of spectral radiances on top of the atmosphere in the thermal infrared spectral region provide a capability to measure atmospheric profiles of temperature and mixing ratios of atmospheric constituents. Another example: targeted measurements of position and velocity of a spacecraft orbiting a planet provide a capability to measure spherical harmonics of the gravity field of this planet.

Keywords Remote sensing · Forward models · Sensitivities

By its basic definition, remote sensing is simply indirect measurement. In a vast variety of practical applications, we cannot measure quantities of interest directly, but we measure some other quantities that are related to quantities of interest by some known relations. For example, targeted measurements of spectral radiances on top of the atmosphere in the thermal infrared spectral region provide a capability to measure atmospheric profiles of temperature and mixing ratios of atmospheric constituents. Another example: targeted measurements of position and velocity of a spacecraft orbiting a planet provide a capability to measure spherical harmonics of the gravity field of this planet.

A mandatory premise of a meaningful remote sensing experiment is the availability of the quantitative framework, which makes it possible to simulate the measured quantities for current estimates of quantities of interest. Such quantitative models are based on available theoretical descriptions of objects under study and processes there, which ultimately produce the quantities that are measured by the remote sensing instruments. Correspondingly, there are two mandatory components of these models: quantitative algorithms, which implement the theoretical description of objects under study, and quantitative description of the measurement procedures by remote sensing instruments.

© The Author(s) 2015
E.A. Ustinov, *Sensitivity Analysis in Remote Sensing*,
SpringerBriefs in Earth Sciences, DOI 10.1007/978-3-319-15841-9_1

In a nutshell, these models provide a quantitative description of the cause—effect relationship between the parameters of interest specifying the object under study and the results of measurements by the remote sensing instruments. In this sense, these models can be termed as *forward models*, because they simulate the corresponding causal link forward: from parameters of interest to measured data.

On the other hand, the very goal of remote sensing is to follow this causal link in the opposite direction: from measured data to parameters of interest. A plain analogy with solution of some, generally non-linear, equation $f(x) = a$ would be instructive here. We have a known function $f(x)$, and we wish to find the value of its argument x, which when substituted in function $f(x)$ causes it to return the value a. In the process of finding a solution for this equation, we follow from the known value a to the argument value x, which yields $f(x) = a$.

A broad and well populated area of the theoretical basis of remote sensing lies in exploring and developing different quantitative methods of retrieving (evaluation of) parameters of interest from measured data. Corresponding, generally nonlinear, inverse problems can be represented in the form analogous to the above equation $f(x) = a$, and methods of their solution are essentially analogous to the Newton-Raphson method of solution of this equation. This is an iterative method, which, at each step, requires evaluation of the function $f(x)$ and its derivative df/dx.

As pointed out above, in remote sensing the causal link between parameters of interest and measured data is followed from effect to cause. This circumstance has profound ramifications, which complicate the practical solution of corresponding inverse problems. In many practical cases, these problems are ill-posed and require application of special methods to stabilize their solutions. Development of these methods began in the middle of the 20th century and is still underway. Numerous monographs on theoretical methods of remote sensing are exclusively devoted to this subject. It is not surprising that there is a popular common perception that theory of remote sensing is reduced to methods of solution of inverse problems.

Of course, this is not true, and this book is intended to demonstrate the opposite. Resorting to the analogy with the Newton-Raphson method of solution of non-linear equations, we see that along with the capability to simulate the measurement data one has to be able to compute derivatives of the simulated data with respect to parameters of interest. In other words, one needs *sensitivities* of simulated data with respect to these parameters.

In principle, if one considers the simulated data as a (scalar, vector, tensor) function of the (scalar, vector, tensor) input parameters, sensitivities are nothing more than corresponding derivatives. But in reality, the situation is more complicated. This book is devoted to methods of efficient computation of these sensitivities.

Chapter 2
Sensitivity Analysis: Differential Calculus of Models

Abstract Models in remote sensing—and in science and engineering, in general—are, essentially, functions of discrete model input parameters, and/or functionals of continuous model input parameters. In this sense, the sensitivities of model output parameters are, essentially, derivatives—partial derivatives with respect to discrete input parameters, and variational derivatives with respect to continuous input parameters. Specificity of models, as compared to functions and functionals in general, is due to the fact that they have two mandatory components. The first component describes the object or process of study, as is. It is a system of a differential equation, or equations, with initial and/or boundary conditions, which is referred to as a *forward problem*. The second component describes the procedure of deriving the output parameters of the model, which simulate the observed quanti-ties, *observables*, from the solution of the forward problem. In contrast to the first component, it is just an analytic expression, which is referred to as an *observables procedure*. In rare practical cases, the forward problem has an analytic solution, and the output parameters are analytic functions or analytic functionals of the input parameters. Correspondingly, analytic evaluation of sensitivities is possible. In most practical cases, only numerical forward solutions are available, and specific approaches of sensitivity analysis considered in this monograph become indispensable.

Keywords Model input and output parameters · Forward problem · Observables · Observables procedure · Sensitivities

2.1 General Considerations

In the context of this book, we define *forward models* as quantitative tools, which return the desired output parameters for given input parameters. These models are assumed to provide an adequate quantitative description of the objects under study, and this description is specified by given values of the input parameters (*model parameters*). The output parameters of these models simulate the results of

© The Author(s) 2015
E.A. Ustinov, *Sensitivity Analysis in Remote Sensing*,
SpringerBriefs in Earth Sciences, DOI 10.1007/978-3-319-15841-9_2

observations (*observables*). Thus, in a nutshell, we deal with functions, whose arguments are the model parameters and the values are the observables.

Indeed, as was pointed out in the Introduction, the forward models used in remote sensing consist of two basic components. The first component, the forward problem, specifies the quantitative description of the object under study as is, based on relevant laws of physics, which govern the spatial structure and temporal behavior of this object. This description—the forward problem—is provided by corresponding differential equations and initial and/or boundary conditions. The second component, the observables procedure, specifies the quantitative recipe of drawing the observables from the solution of the forward problem—the *forward solution*. This recipe is provided in the form of a closed-form analytic expression, which converts the forward solution into the observables.

Thus, abstracting from the inner workings of the forward models, they represent essentially functions whose arguments are the model parameters and values are the observables. There is a caveat though. In many practical cases, the models involve the continuous parameters, which are functions themselves—functions of space and/or time. Then the forward models become the functionals defined on these functions. From the viewpoint of practical implementation, there is little or no difference between functionals and functions of many variables, because the practical implementation requires a representation of those functions on an adequate grid of their arguments. But, as we will see in relevant examples below, from the viewpoint of analytic work that is necessary to conduct the sensitivity analysis in each specific case, it is instructive to treat the models with continuous parameters as functionals, and to apply corresponding tools of variational analysis.

In a relatively small number of practical cases, the forward problems used in practical forward models of remote sensing have analytic solutions. For example, this is the case in remote sensing of planetary atmospheres in the thermal spectral region, when atmospheric scattering can be neglected (see Sects. 4.3 and 5.3). In such cases, the forward solution can be represented as a closed-form analytic expression, which after the application of the observables procedure results in an analytic expression of observables directly through the model input parameters. Then, depending upon whether the given input parameter is just a constant or a function of space and/or time variables, the observables are functions or functionals. Accordingly, the sensitivities can be found using standard techniques of the differential calculus or variational calculus.

In most practical cases, though, the corresponding forward problems can be solved only numerically. This means that for given numerical values of the input parameters, the forward solution is obtained using appropriate numerical methods. This means that the analytic relation between the model input parameters and observables does not exist. This is where the methods of sensitivity analysis become indispensable. At this point we need to take a more detailed look at sensitivities with respect to different types of model parameters and observables.

2.2 Input and Output Parameters of Models

Without losing any generality, the model input parameters can be divided into two broad groups: discrete parameters and continuous parameters. Discrete parameters are constants, which do not depend on any arguments, such as space, time, etc. Continuous parameters are functions with essentially the same domain of arguments, as the forward solutions. Consider a couple of simplified examples.

Motion of a material point in the planetary gravity field. If the finite size of the material object in the gravity field can be neglected as compared to the scale of spatial variation of the gravity field of the planet, then this object can be considered as a material point, and its motion, assumed here to be non-relativistic, is described by a forward problem in the form:

$$\begin{cases} \dfrac{d^2\mathbf{r}}{dt^2} = \mathbf{g}(\mathbf{r}, t) \\ \left. \dfrac{d\mathbf{r}}{dt}\right|_{t=0} = \mathbf{v}_0 \\ \mathbf{r}|_{t=0} = \mathbf{r}_0 \end{cases} \tag{2.1}$$

The model input parameters here consist of two discrete parameters, $\mathbf{v}_0$ and $\mathbf{r}_0$, and one continuous parameter $\mathbf{g}(\mathbf{r}, t)$. The forward solution $\mathbf{r}(t)$ is a function of time t.

Transfer of thermal radiation in the non-scattering planetary atmosphere. Neglecting the vertical span of the planetary atmosphere as compared to the radius of the planet, the radiative transfer in the non-scattering atmosphere can be described by a forward problem in the form:

$$\begin{cases} u\dfrac{dI}{dz} + \alpha(z)I(z, u) = \alpha(z)B(z) \\ I(z, u) = 0, \quad \text{for } z = 0, u > 0 \\ I(z, u) = 2A \int_0^1 I(z, u')\, u'du' + B_s, \quad \text{for } z = z_0, u < 0 \end{cases} \tag{2.2}$$

Here we have two discrete parameters, albedo A and source function B_s of the underlying surface, and two continuous parameters, extinction coefficient $\alpha(z)$, and source function $B(z)$ of the atmosphere itself. The forward solution $I(z, u)$ is a function of the vertical coordinate z and of cosine u of the nadir angle of propagation.

The practical implementation of the retrieval algorithms in the form of computer programs intended for the interpretation of practical data involves the representation of the continuous parameters as finite-dimensional arrays with values corresponding to an appropriate grid of argument values of those continuous parameters. But in many practical cases, the analytic work, which is necessary to be done in order to derive the analytic background of the retrieval algorithms, is easier to perform in terms of continuous parameters considered as functions, using appropriate tools of variational analysis. This will be demonstrated in the sections to follow.

On the other hand, the output parameters, the observables R, are always discrete parameters due to the nature of measurements themselves. In the first example above, such observables may be relative distances and velocities of the material point as measured from some known location of the observer. At each instant of observation:

$$\mathbf{R} = |\mathbf{r} - \mathbf{r}_i|$$
$$\mathbf{S} = |\mathbf{V} - \mathbf{V}_i| \tag{2.3}$$

In practice, the measured values are always integrated over some finite span of arguments of the forward solution: time, space, viewing angle, etc. But the integration results are always benchmarked by some instant values of these arguments. More information will be provided in the sections to follow.

2.3 Sensitivities: Just Derivatives of Output Parameters with Respect to Input Parameters

As mentioned in the Introduction, sensitivity of any output parameter with respect to any input parameter is merely a derivative of a suitable type. Sensitivity of the observable R with respect to a discrete input parameter p isa partial derivative:

$$K = \frac{\partial R}{\partial p} \tag{2.4}$$

Sensitivity of the observable R with respect to a continuous input parameter $p(x)$ is a variational derivative:

$$K(x) = \frac{\delta R}{\delta p(x)} \tag{2.5}$$

There are a few different definitions of the variational derivative, which sometimes is also called the functional derivative. For all practical purposes in this book, the variational derivative with respect to the continuous parameter $p(x)$ is defined as a kernel of the linear integral expression

$$\delta_p R = \int_{D_x} \frac{\delta R}{\delta p(x)} \delta p(x) \, \mathrm{d}x \tag{2.6}$$

Here, the integration is conducted over the domain D_x of arguments x of the parameter $p(x)$; $\delta p(x)$ is the variation of the input parameter $p(x)$, and $\delta_p R$ is the variation of the output parameter R caused by the variation $\delta p(x)$.

For a continuous parameter specified on a grid of its arguments, there exists a simple relationship between values of the variational derivative on this grid and values of partial derivatives with respect to grid values of the continuous parameter. The accuracy of this relationship depends on the mesh width of this grid. As an example, assume that $p(x)$ is defined on an interval $x \in [a, b]$ represented by a set of grid values $p_j = p(x_j)$, $(j = 1, \ldots n)$. Then, the output parameter R becomes a function of n variables $\{x_j\}$, and Eq. (2.6) can be approximated in the form:

$$\delta_p R = \int_{D_x} \frac{\delta R}{\delta p(x)} \delta p(x) \, dx \approx \sum_j \left(\frac{\delta R}{\delta p(x)} \right)\bigg|_{x_j} \delta p_j \Delta_j x \qquad (2.7)$$

From Eq. (2.7) we immediately obtain:

$$\frac{\partial R}{\partial p_j} \approx \left(\frac{\delta R}{\delta p(x)} \right)\bigg|_{x_j} \Delta_j x, \text{ or vice versa, } \left(\frac{\delta R}{\delta p(x)} \right)\bigg|_{x_j} \approx \frac{1}{\Delta_j x} \frac{\partial R}{\partial p_j} \qquad (2.8)$$

The definition of variational derivative, Eq. (2.6), has an important practical value. Throughout this book we will derive the expressions for sensitivities to continuous parameters—which are variational derivatives—by transforming the expressions for variations of the output parameters to the form of Eq. (2.6) and obtaining the sensitivities as kernels of resulting integral expressions. Equations (2.5) and (2.6) will serve as a definition of sensitivities to continuous parameters.

Consider a special kind of a functional: the value of a function $f(\xi)$ defined on the interval $[a, b]$ at the specified value of its argument $\xi = x$. Representing $f(x)$ in the form

$$f(x) = \int_a^b \delta(\xi - x) f(\xi) \, d\xi \qquad (2.9)$$

we take the variations of both sides of Eq. (2.9):

$$\delta f(x) = \int_a^b \delta(\xi - x) \delta f(\xi) \, d\xi \qquad (2.10)$$

Comparing Eq. (2.10) with the definition of variational derivative, Eqs. (2.5) and (2.6), we obtain:

$$\frac{\delta f(x)}{\delta f(\xi)} = \delta(\xi - x) \tag{2.11}$$

In a similar fashion, one can derive an expression for the variational derivative of the ordinary derivative of the function $f'(x)$ with respect to the function $f(x)$ itself:

$$\frac{\delta f'(x)}{\delta f(\xi)} = -\delta'(\xi - x) \tag{2.12}$$

where $\delta'(x) = \mathrm{d}\delta(x)/\mathrm{d}x$ is the derivative of the δ-function. We have:

$$\delta f'(x) = \int_a^b \delta(\xi - x)\delta f(\xi)\,\mathrm{d}\xi \tag{2.13}$$

Integrating by parts we obtain:

$$\delta f'(x) = \delta(\xi - x)\delta f(\xi)\big|_a^b - \int_a^b \delta'(\xi - x)\delta f(\xi)\,\mathrm{d}\xi \tag{2.14}$$

The off-integral term in Eq. (2.14) equals zero for all values of x within the interval $[a, b]$. Comparing the resulting Eq. (2.14) with the definition of variational derivative, Eqs. (2.5) and (2.6), we see that Eq. (2.12) is valid everywhere within this interval.

In a number of applications considered in chapters that follow, we will need to convert the linear variational equation

$$L\delta X = \delta S \tag{2.15}$$

into an equation for corresponding variational derivatives with respect to some continuous parameter:

$$L\frac{\delta X}{\delta a} = \frac{\delta S}{\delta a} \tag{2.16}$$

Here, δa is the variation of the parameter $a(x)$, which results in a corresponding variation of the right-hand term $S(x)$, which in turn, results in a corresponding variation of the solution $X(x)$. The linear operator may, in general, be a function of x, too.

In the applications below, at each given value of the argument $x = \xi$, the function $S(x)$ depends on the parameter $a(x)$ only at the same value of the argument $x = \xi$. This means that S is a function of $a(x)$, not a functional of $a(\xi)$. Accordingly,

$$\delta S(x) = \int_a^b \delta(\xi - x) \cdot \frac{\partial S(\xi)}{\partial a(\xi)} \cdot \delta a(\xi) \, d\xi = \int_a^b \delta(\xi - x) \cdot \frac{\partial S(x)}{\partial a(x)} \cdot \delta a(\xi) \, d\xi \quad (2.17)$$

Comparing with the definition of the variational derivative, Eqs. (2.5) and (2.6), we have:

$$\frac{\delta S(x)}{\delta a(\xi)} = \delta(\xi - x) \cdot \frac{\partial S(x)}{\partial a(x)} \quad (2.18)$$

Thus, the right-hand term of Eq. (2.12) has the form:

$$\delta S(x) = \frac{\partial S(x)}{\partial a(x)} \delta a(x) \quad (2.19)$$

On the other hand, the solution $\delta X(x)$ of Eq. (2.15) depends on the variation of the parameter $a(\xi)$ in the whole interval $\xi \in [x_0, x_1]$. In other words, $X(x)$ is a functional of $a(\xi)$, while still being a function of x:

$$X(x) = X[a(\xi), x] \quad (2.20)$$

Thus, in the variational equation, Eq. (2.15)

$$\delta X(x) = \int_a^b \frac{\delta X(x)}{\delta a(\xi)} \delta a(\xi) d\xi \quad (2.21)$$

Substituting Eqs. (2.17) and (2.21) in Eq. (2.15) and moving the right-hand term into the left side, we have:

$$L \int_a^b \frac{\delta X(x)}{\delta a(\xi)} \delta a(\xi) d\xi - \int_a^b \delta(\xi - x) \cdot \frac{\partial S(x)}{\partial a(x)} \cdot \delta a(\xi) d\xi = 0 \quad (2.22)$$

Recalling that the operator L may be a function of x, and observing that x is not an integration variable in Eq. (2.22), we can rewrite this equation as

$$\int_a^b \left(L \frac{\delta X(x)}{\delta a(\xi)} - \delta(\xi - x) \cdot \frac{\partial S(x)}{\partial a(x)} \right) \delta a(\xi) d\xi = 0 \quad (2.23)$$

Finally, demanding that Eq. (2.23) be satisfied for an arbitrary variation δa, we obtain the equation for variational derivatives, Eq. (2.16), which, in a detailed form can be written as

$$L\frac{\delta X(x)}{\delta a(\xi)} = \delta(\xi - x) \cdot \frac{\partial S(x)}{\partial a(x)} \qquad (2.24)$$

We will use this result in applications of the linearization approach of sensitivity analysis to various forward problems considered in the chapters that follow.

Chapter 3
Three Approaches to Sensitivity Analysis of Models

Abstract There are three ways to implement the sensitivity analysis of quantitative models. The simplest finite-difference (FD) approach requires multiple re-runs of the forward model according to the number of model input parameters. Although this approach uses the forward model without any modifications and does not require any analytic work, its application results in a very computer-intensive algorithm, which may become a prohibitive factor in practical applications to models with a large number of input parameters. Two other approaches of sensitivity analysis—the linearization approach and adjoint approach—are substantially more computer-efficient and require just single runs of a corresponding model derived from the initial, baseline model. The general formulation and comparison of these approaches is presented in this chapter.

Keywords Sensitivity analysis · Finite-difference approach · Linearization approach · Adjoint approach

3.1 Finite-Difference Approach

In high-level notations, the forward problem can be written in the form of a generally non-linear operator equation

$$N[X] = S \tag{3.1}$$

where the operator N combines all operations on the forward solution X, and the right-hand term S combines all terms that do not include X. All model parameters are contained either in the operator N or in the right-hand term S, both of which are some functions of these parameters. Using similar shorthand notations, computation of observables R from the forward solution X can be represented in the form:

$$R = M[X] \tag{3.2}$$

© The Author(s) 2015

E.A. Ustinov, *Sensitivity Analysis in Remote Sensing*,

SpringerBriefs in Earth Sciences, DOI 10.1007/978-3-319-15841-9_3

where M is, in general, a non-linear functional describing the procedure of modeling the observables.

To compute the sensitivities of observables R to model input parameters, each input parameter p, one by one, is varied by a suitable small amount Δp, and the resulting forward solution $\tilde{X}$ is used to obtain the varied observables $\tilde{R}$, and corresponding sensitivities are obtained as ratios

$$\frac{\partial R}{\partial p} \approx \frac{\tilde{R} - R}{\Delta p} \tag{3.3}$$

Since this process has to be repeated for each input parameter, the required computer time increases (as compared to the single run of the forward model) by a factor equal to the number of model input parameters.

3.2 Linearization Approach

Following this approach, the forward model Eq. (3.1) is linearized around its known solution X, hereafter referred to as the *baseline solution*. We assume that the model parameters experience small perturbations. Then, the left-hand and right-hand terms of Eq. (3.1) experience variations $\delta(N[X])$ and δS correspondingly, and we have:

$$\delta(N[X]) = \delta S \tag{3.4}$$

The variation $\delta(N[X])$ is, further on, represented in the form

$$\delta(N[X]) = \delta N[X] + L\delta X \tag{3.5}$$

Here, the linear operator L represents a linearization of the operator N around the baseline solution X.

As an example, consider the non-linear operator $N[X]$ in the form:

$$N[X] = a(x)\frac{dX}{dx} + b(x)X^2(x) \tag{3.6}$$

Its variation

$$\delta(N[X]) = a(x)\frac{d\delta X}{dx} + \delta a(x)\frac{dX}{dx} + 2b(x)X(x)\delta X(x) + \delta b(x)X^2(x) \tag{3.7}$$

can be represented in the form of Eq. (3.5), where

$$\delta N[X] = \delta a(x)\frac{dX}{dx} + \delta b(x)X^2(x) \tag{3.8}$$

and

$$L\delta X = a(x)\frac{d\delta X}{dx} + 2b(x)X(x)\delta X(x) \tag{3.9}$$

Substituting Eq. (3.5) in Eq. (3.4) and re-arranging the terms in the resulting equation we obtain

$$L\delta X = \delta S - \delta N[X] \tag{3.10}$$

Note that only the right-hand term of the linearized forward problem Eq. (3.21) depends on variations of model parameters; the operator L of the linearized forward problem Eq. (3.10) remains the same. This provides substantial savings of computer time, as compared to the finite-difference approach.

For discrete model parameters p_j, representing variations in Eq. (3.10) in the form:

$$\delta\bullet = \frac{\partial\bullet}{\partial p_j}\delta p_j \tag{3.11}$$

we obtain the linearized forward problem for corresponding sensitivities of the forward solution $X(x)$:

$$L\frac{\partial X}{\partial p_j} = \frac{\partial S}{\partial p_j} - \frac{\partial N[X]}{\partial p_j} \tag{3.12}$$

For continuous model parameters $p(x)$, representing variations in Eq. (3.10) in the form:

$$\delta\bullet(x) = \int_{Dx}\frac{\delta\bullet(x)}{\delta p(\xi)}\delta(\xi)d\xi = \delta(\xi - x)\cdot\frac{\partial\bullet}{\partial p(x)} \tag{3.13}$$

we obtain the linearized forward problem for corresponding sensitivities of the forward solution $X(x)$:

$$L\frac{\delta X}{\delta p(\xi)} = \delta(\xi - x)\left(\frac{\partial S}{\partial p(\xi)} - \frac{\partial N[X]}{\partial p(\xi)}\right) \tag{3.14}$$

Further on, the observables procedure Eq. (3.2) is linearized with respect to the forward solution X, resulting into a linear functional, which can be written in the form of an inner product

$$\delta R = (W, \delta X) = \int_{D_x} W(x)\delta X(x)dx \tag{3.15}$$

This yields resulting expressions for sensitivities of observables R to discrete model parameters:

$$\frac{\partial R}{\partial p_j} = \left(W, \frac{\partial X}{\partial p_j} \right), \tag{3.16}$$

and continuous model parameters:

$$\frac{\delta R}{\delta p(\xi)} = \left(W, \frac{\delta X}{\delta p(\xi)} \right) \tag{3.17}$$

3.3 Adjoint Approach

The adjoint approach is based on the use of the operator L^*, adjoint to the linear operator L of the linearized forward problem Eq. (3.10). Rewriting this problem as

$$LX' = S' \tag{3.18}$$

where $S' = \delta S - \delta N[X]$ and $X' = \delta X$, and rewriting the linearized observables procedure Eq. (3.15) as

$$R' = (W, X') \tag{3.19}$$

this approach can be outlined as follows. By definition, the adjoint operator L^* satisfies the Lagrange identity

$$(g, Lf) = (L^* g, f) \tag{3.20}$$

for an arbitrary pair of functions f and g in the domain of L. Then, the solution X^* of the adjoint problem

$$L^* X^* = W \tag{3.21}$$

provides an alternative way to compute the linearized observables:

$$R' = (X^*, S') \tag{3.22}$$

This can be demonstrated by multiplying the adjoint problem Eq. (3.21) by X' and the linearized forward problem Eq. (3.18) by X^*, and then comparing left and right sides of the resulting equalities:

$$(L^* X^*, X') = (W, X'), \quad (X^*, LX') = (X^*, S') \tag{3.23}$$

Since the left-hand terms (L^*X^*, X') and (X^*, LX') are equal by definition of the adjoint operator, the right-hand terms (W, X') and (X^*, S') are also equal. Then, Eq. (3.22) immediately follows from Eq. (3.19).

Now, returning from the abbreviated Eq. (3.18) to Eq.(3.10) and replacing in Eq. (3.22) $R' \to \delta R$ and $S' \to \delta S - \delta N[X]$, we obtain a direct expression of the variation of the observables through variations of the operator $N[X]$ and the right-hand term S of the forward problem Eq. (3.1):

$$\delta R = (X^*, \delta S - \delta N[X]) \tag{3.24}$$

This yields expressions for sensitivities of observables to discrete and continuous parameters in the form:

$$\frac{\partial R}{\partial p_j} = \left(X^*, \frac{\partial S}{\partial p_j} - \frac{\partial N[X]}{\partial p_j} \right), \tag{3.25}$$

and

$$\frac{\delta R}{\delta p(\xi)} = \left(X^*, \frac{\delta S}{\delta p(\xi)} - \frac{\delta N[X]}{\delta p(\xi)} \right) \tag{3.26}$$

3.4 Comparison of Three Approaches

As pointed out above, the finite-difference (FD) approach is the simplest one to implement. No analytic work is necessary, one needs just to add a couple of additional loops in the computer code—to run the baseline forward problem for individual varied input parameters, and to feed resulting finite-difference approximations of resulting sensitivities into the corresponding rows of the jacobian matrix. Computing time, as compared to a single run of the baseline forward problem increases by a factor corresponding to the number of input parameters of interest, which is a significant drawback of the FD approach. On the other hand, the efficiency of computers is still rapidly increasing, year by year, and this trend will continue years and years ahead. This is why the FD approach will remain a viable option for many problems, at least in the near future.

The linearization approach is much more efficient than the FD approach. As we have seen above, the operator L in the left side of Eq. (3.12) or Eq. (3.14) is the same for all the sensitivities of interest, and only the right-hand terms are different for different model parameters. The practice of implementing algorithms to solve the differential equations Partial differential equation tells us that most of the computations and intermediate data flows are associated with the terms in the left side of differential equations and of their initial and/or boundary conditions. The overhead due to right-hand terms is relatively small. On the other hand, the implementation of the linearization approach needs some additional analytic effort.

One has to linearize the differential equations and initial and/or boundary conditions of the baseline forward problem. The good news here is that the resulting linearized forward problem can be solved with the same method that is used to solve its parent, the baseline forward problem, and development of the corresponding computer code is relatively simple.

The adjoint approach becomes more efficient in comparison with the linearization approach when the number of model input parameters exceeds the number of the observables. As in the case of the linearization approach, the adjoint operator L^* is the same for all observables, and only the right-hand terms are different for different observables. On the other hand, the implementation of the adjoint approach needs even more additional analytic effort. But, as the linearized forward problem, the adjoint problem can be solved with the same method, which is applied to the baseline forward problem. This is because the operations in the left-hand terms of the adjoint problem are very similar to those of the linearized forward problem. This will be illustrated by the examples given in the following chapters.

The adjoint approach has definite advantage over the linearization approach, when the number of model parameters substantially exceeds the number of observables. This is especially the case, when some, or all model parameters are continuous and have to be specified on a grid of their arguments. Then the number of separate grid values of model parameters, and, correspondingly, the number of different right-hand terms in the linearized forward problem may easily be of the order of many tens, or even hundreds. Typical examples are profiles of temperature and composition in planetary atmospheres, or profiles of mass density and bulk modulus and shear modulus in planetary interiors. In cases like these, the adjoint approach is definitely preferable.

In the remainder of this book we will consider applications of the linearization and adjoint approaches to various types of forward models.

References

It is difficult to trace down the first publications on applications of the finite-difference and linearization approaches in remote sensing. As for the adjoint approach, to the best of author's knowledge, the pioneering paper on its application in remote sensing is listed in reference.

Marchuk GM (1964) Equation for the value of information from weather satellites and formulation of inverse problems. Cosmic Res 2:394–409

Chapter 4
Sensitivity Analysis of Analytic Models: Applications of Differential and Variational Calculus

Abstract Most forward problems in remote sensing can be solved only using numerical methods. Correspondingly, the resulting models are by necessity numerical ones. Nevertheless, in some practical cases corresponding forward problems can be solved analytically, and resulting models become analytical. In such cases, sensitivities to discrete and continuous input parameters can be found using the conventional methods of differential and variational calculus. In this chapter we consider the sensitivity analysis of analytic models. For demonstration purposes, we first consider two simple demo models. Then a realistic model will be considered, which is actually used in practice of remote sensing of the Earth and planetary atmospheres.

Keywords Analytic models · Differential calculus · Variational calculus

4.1 Linear Demo Model

Consider the forward problem for this model in the form:

$$\begin{cases} \dfrac{dX}{dt} + a(t)X(t) = 0 \\ X(t_0) = X_0 \end{cases} \tag{4.1}$$

Here, the continuum input parameter $a(t)$ and forward solution $X(t)$ are functions of a single argument, time t. The right-hand term X_0 of the initial-value condition is a discrete input parameter. We assume that the forward problem Eq. (4.1) is integrated over the interval $[t_0, t_1]$, and as an observable R we will use the value of the forward solution at the instant t_1:

$$R = X(t_1) \tag{4.2}$$

© The Author(s) 2015
E.A. Ustinov, *Sensitivity Analysis in Remote Sensing*,
SpringerBriefs in Earth Sciences, DOI 10.1007/978-3-319-15841-9_4

Solution of the differential equation in Eq. (4.1) can be found by rewriting it in the form:

$$\frac{dX}{X(t)} = -a(t)\ dt, \quad \text{or} \quad d\ln X(t) = -a(t)\ dt$$

After integration and accounting for the initial-value condition in Eq. (4.1) we obtain the forward solution $X(t)$ in the form:

$$X(t) = X_0 \exp[-A(t)] \tag{4.3}$$

where

$$A(t) = \int_{t_0}^{t} a(t)\ dt \tag{4.4}$$

is the primitive function of $a(t)$, so that its derivative $dA/dt = a(t)$.

The observable, Eq. (4.2) takes the form:

$$R = X_0 \exp[-A(t_1)] \tag{4.5}$$

Its sensitivity to the discrete parameter X_0 can be found by taking corresponding partial derivative of Eq. (4.5). We have:

$$\frac{\partial R}{\partial X_0} = \exp[-A(t_1)] = \frac{R}{X_0} \tag{4.6}$$

Sensitivity of the observable R to the continuous parameter $a(t)$ can be found by linearizing Eq. (4.5) with respect to $a(t)$ and expressing the corresponding variation of the observable $\delta_a R$ through the variation $\delta a(t)$:

$$\delta_a R = \delta_a \left[X_0 \exp\left(-\int_{t_0}^{t_1} a(t)\ dt \right) \right] = X_0 \exp[-A(t_1)] \cdot \left(-\int_{t_0}^{t_1} \delta a(t)\ dt \right)$$

$$= -R \int_{t_0}^{t_1} \delta a(t)\ dt$$

Comparing with the general definition of sensitivity to the continuous parameter, Eq. (2.6), we obtain:

$$\frac{\delta R}{\delta a(t)} = -R \tag{4.7}$$

As a form of validation, this expression will be obtained again in the Chap. 5 using two alternative approaches of sensitivity analysis—linearization and adjoint approaches described in the Chap. 3.

4.2 Non-linear Demo Model

Consider the forward problem for this model in the form:

$$\begin{cases} \dfrac{dX}{dt} + a(t)X^2(t) = 0 \\ X(t_0) = X_0 \end{cases} \tag{4.8}$$

As with the linear demo model, we have one continuum input parameter $a(t)$ and one discrete input parameter X_0. We also assume that the forward problem Eq. (4.8) is integrated over the interval $[t_0, t_1]$, and the observable R has the same form as in Eq. (4.2).

Solution of the differential equation in Eq. (4.8) can be found by rewriting it in the form:

$$\frac{dX}{X^2(t)} = -a(t)\,dt, \quad \text{or} \quad d\left[\frac{1}{X(t)}\right] = a(t)\,dt$$

After integration and accounting for the initial-value condition in Eq. (4.8) we obtain the forward solution $X(t)$ in the form:

$$X(t) = \left(A(t) + X_0^{-1}\right)^{-1} \tag{4.9}$$

The observable, Eq. (4.2) takes the form:

$$R = \left(A(t_1) + X_0^{-1}\right)^{-1} \tag{4.10}$$

Its sensitivity to the discrete parameter X_0 is obtained by taking the corresponding partial derivative of Eq. (4.10). We have:

$$\frac{\partial R}{\partial X_0} = \frac{\partial}{\partial X_0}\left[\left(A(t_1) + X_0^{-1}\right)^{-1}\right] = -\left(A(t_1) + X_0^{-1}\right)^{-2}\cdot\left(-X_0^{-2}\right) = \left(\frac{R}{X_0}\right)^2 \tag{4.11}$$

Sensitivity of the observable R to the continuous parameter $a(t)$ can be found by linearizing Eq. (4.10):

$$\delta_a R = \delta_a \left[\left(A(t_1) + X_0^{-1} \right)^{-1} \right] = - \left(A(t_1) + X_0^{-1} \right)^{-2} \cdot \int_{t_0}^{t_1} \delta a(t) \, dt = -R^2 \int_{t_0}^{t_1} \delta a(t) \, dt$$

Comparing with the general definition of sensitivity to the continuous parameter, Eq. (2.6), we obtain:

$$\frac{\delta R}{\delta a(t)} = -R^2 \tag{4.12}$$

This expression will be obtained again in the Chap. 5 using two alternative approaches of sensitivity analysis—the linearization and adjoint approaches described in Chap. 3.

4.3 Model of Radiances of a Non-scattering Planetary Atmosphere

This model is widely used in remote sensing of non-scattering planetary atmospheres in the thermal spectral region of electromagnetic (EM) radiation. The forward problem is a classic radiative transfer (RT) problem, which for a plane-parallel model of the atmosphere can be written in the form:

$$\begin{cases} u \dfrac{dI}{dz} + \kappa(z)I(z,u) = \kappa(z)B(z) \\ I(0,u) = 0; \quad u > 0 \\ I(z_0,u) = \varepsilon B_s; \quad u < 0 \end{cases} \tag{4.13}$$

Here $I(z,u)$ is the intensity of thermal EM radiation depending on the vertical coordinate in the atmosphere z measured from the top of the atmosphere (TOA) and cosine u of the nadir angle of direction of propagation of EM radiation, $\kappa(z)$ is the absorption coefficient, $B(z)$ is the intensity of source EM radiation in the planetary atmosphere, ε is the emissivity of the planetary surface, and B_s is the intensity of source EM radiation emanating from the planetary surface. Note that solutions of the problem Eq. (4.34) for $u > 0$ and $u < 0$ are essentially disengaged, with the upper boundary condition at $z = 0$ serving as initial-value condition for $I(z, u > 0)$, and the lower boundary condition at $z = z_0$ serving as initial-value condition for $I(z, u < 0)$.

The forward problem, Eq. (4.13), has two continuous model parameters: $\kappa(z)$, $B(z)$, and two discrete model parameters: ε, B_s. The observable is intensity of EM radiation at TOA observed at cosine μ of the nadir angle of observation

$$R(\mu) = I(0, -\mu) \tag{4.14}$$

and thus is drawn from the solution $I(z, u < 0)$.

The differential equation of the forward RT problem, Eq. (4.13) has a form of an ordinary differential equation (ODE) of first order with variable coefficients:

$$y' + p(x)y(x) = q(x) \tag{4.15}$$

Its solution is obtained in a standard way, by multiplying Eq. (4.15) by an auxiliary function $M(x)$, which satisfies the condition $M' = M(x)p(x)$ and thus transforms the left side of Eq. (4.15) into a derivative of the product $M(x)y(x)$. Integrating this condition on $M(x)$ we have:

$$M(x) = \exp\left(\int p(x)\, dx\right) \tag{4.16}$$

and integrating the transformed Eq. (4.15) we obtain:

$$y(x) = \frac{\int M(x)q(x)\, dx + C}{M(x)} \tag{4.17}$$

Note that the observable Eq. (4.14) is drawn from the solution $I(z, u < 0)$ integrated from the lower boundary condition of the forward problem Eq. (4.13). Replacing $p(x) \to \kappa(z)/u$, $q(x) \to \kappa(z)B(z)/u$ we obtain:

$$I(z, u) = -\frac{1}{u}\int_{z}^{z_0} B(z') \exp\left(\frac{1}{u}\int_{z}^{z'} \kappa(z'')dz''\right) \kappa(z')dz' + \varepsilon B_s \exp\left(\frac{1}{u}\int_{z}^{z_0} \kappa(z')dz'\right) \tag{4.18}$$

Substituting in Eq. (4.14) we obtain the observable $R(\mu)$ in the form:

$$R(\mu) = \frac{1}{\mu}\int_{0}^{z_0} B(z)t(z, \mu)\kappa(z)dz + t(z_0, \mu)\varepsilon B_s \tag{4.19}$$

where

$$t(z, \mu) = \exp\left(-\frac{1}{\mu}\int_{0}^{z} \kappa(z')dz'\right) \tag{4.20}$$

is atmospheric transmittance from level z to TOA.

Sensitivity of the observable Eq. (4.19) to the discrete parameters ε and B_s can be immediately found by taking corresponding partial derivatives. We have:

$$\frac{\partial R(\mu)}{\partial B_s} = t(z_0, \mu)\,\varepsilon \tag{4.21}$$

$$\frac{\partial R(\mu)}{\partial \varepsilon} = t(z_0, \mu)B_s \tag{4.22}$$

Sensitivity of the observable $R(\mu)$ to the continuous parameter $B(z)$ can be found by linearizing Eq. (4.19) with respect to $B(z)$:

$$\delta_B R(\mu) = \delta_B \left(\frac{1}{\mu} \int_0^{z_0} B(z)t(z,\mu)\kappa(z)\mathrm{d}z + t(z_0,\mu)\varepsilon B_s \right) = \frac{1}{\mu} \int_0^{z_0} \delta B(z)t(z,\mu)\kappa(z)\mathrm{d}z$$

Comparing with the general definition of the sensitivity to the continuous parameter, Eq. (2.6), we obtain:

$$\frac{\delta R(\mu)}{\delta B(z)} = \frac{1}{\mu}\kappa(z)t(z,\mu) \tag{4.23}$$

To obtain the sensitivity to the absorption coefficient $\kappa(z)$, we need to transform Eq. (4.19) to a more suitable expression. We have:

$$R(\mu) = \frac{1}{\mu} \int_0^{z_0} B(z)t(z,\mu)\kappa(z)\mathrm{d}z + t(z_0,\mu)\varepsilon B_s = -\int_0^{z_0} B(z)\mathrm{d}t(z,\mu) + t(z_0,\mu)\varepsilon B_s$$

After integration by parts we obtain

$$R(\mu) = B(0) + \int_0^{z_0} t(z,\mu)\mathrm{d}B(z) + t(z_0,\mu)[\varepsilon B_s - B(z_0)] \tag{4.24}$$

Linearizing Eq. (4.24) with respect to $t(z,\mu)$ we have:

$$\delta_t R(\mu) = \int_0^{z_0} \delta t(z,\mu)\mathrm{d}B(z) + \delta t(z_0,\mu)[\varepsilon B_s - B(z_0)] \tag{4.25}$$

Further on, linearizing the atmospheric transmittance Eq. (4.20) with respect to $\kappa(z)$, we have:

$$\delta_\kappa t(z,\mu) = \exp\left(-\frac{1}{\mu}\int_0^z \kappa(z')dz'\right)\cdot\left(-\frac{1}{\mu}\int_0^z \delta\kappa(z')dz'\right) = -\frac{1}{\mu}t(z,\mu)\int_0^z \delta\kappa(z')dz'$$

$$(4.26)$$

Substituting Eq. (4.26) in Eq. (4.25) and changing the order of integration in the resulting double integration term we have:

$$\delta_\kappa R(\mu) = -\frac{1}{\mu}\int_0^{z_0} dz\delta\kappa(z)\left\{\int_z^{z_0} t(z',\mu)dB(z') + t(z_0,\mu)[\varepsilon B_s - B(z_0)]\right\} \quad (4.27)$$

Comparing with the general definition of the sensitivity to the continuous parameter, Eq. (2.6), we obtain:

$$\frac{\delta R(\mu)}{\delta\kappa(z)} = -\frac{1}{\mu}\left\{\int_z^{z_0} t(z',\mu)dB(z') + t(z_0,\mu)[\varepsilon B_s - B(z_0)]\right\} \quad (4.28)$$

Using integration by parts, Eq. (4.28) can be rewritten in a more compact form:

$$\frac{\delta R(\mu)}{\delta\kappa(z)} = -\frac{1}{\mu}[r(z,\mu) - t(z,\mu)B(z)] \quad (4.29)$$

where the upwelling radiance

$$r(z,\mu) = \int_{z_0}^z B(z')dt(z',\mu) + t(z_0,\mu)\varepsilon B_s \quad (4.30)$$

is a byproduct of integration of radiances via Eq. (4.19):

$$R(\mu) = r(0,\mu) = \int_{z_0}^0 B(z)dt(z,\mu) + t(z_0,\mu)\varepsilon B_s \quad (4.31)$$

In the next chapter, the results derived above using direct linearization will be obtained using the linearization and adjoint approaches.

It is worth pointing out here that the atmospheric Planck function $B(z)$ and surface Planck function B_s are not the input parameters of interest per se. Instead, of interest are the atmospheric temperature $T(z)$ and surface temperature T_s. Corresponding sensitivities can be obtained from Eqs. (4.23) and (4.21):

$$\frac{\delta R(\mu)}{\delta T(z)} = \frac{\delta R(\mu)}{\delta B(z)} \cdot \frac{\partial B}{\partial T}\bigg|_{T(z)} = \frac{1}{\mu} \kappa(z) t(z, \mu) \frac{\partial B}{\partial T}\bigg|_{T(z)} \tag{4.32}$$

and

$$\frac{\partial R(\mu)}{\partial T_s} = \frac{\partial R(\mu)}{\partial B_s} \cdot \frac{\partial B}{\partial T}\bigg|_{T_s} = \varepsilon t(z_0, \mu) \frac{\partial B}{\partial T}\bigg|_{T_s} \tag{4.33}$$

where temperature derivatives of the Planck function are available analytically.

Similarly, the absorption coefficient $\kappa(z)$ is not the input parameter of interest per se. Instead, of interest are the vertical profiles of atmospheric constituents, which contribute to the atmospheric absorption. As an example, let's assume that we wish to evaluate the sensitivity of the observed radiances $R(\mu)$ to the vertical profile of some minor gaseous atmospheric constituent, which contributes to atmospheric opacity at frequencies, where these radiances are observed. The absorption coefficient of this constituent $\kappa_m(z)$ can be represented in the form:

$$\kappa_m(z) = \sigma_m(z) f_m(z) n_0(z) \tag{4.34}$$

where $\sigma_m(z), f_m(z)$ are the absorption cross section and mixing ratio of this gaseous constituent respectively, and $n_0(z)$ is the total number density of the atmosphere. Variation of the total absorption coefficient $\kappa(z)$ due to variation of $f_m(z)$ can be presented in the form:

$$\delta \kappa(z) = \delta \kappa_m(z) = \kappa_m(z) \delta \ln \kappa_m(z) = \kappa_m(z) \delta \ln f_m(z) \tag{4.35}$$

Accordingly, the corresponding variation of radiance $R(\mu)$ can be written in the form:

$$\delta R(\mu) = \int\limits_0^{z_0} \frac{\delta R(\mu)}{\delta \kappa(z)} \delta \kappa(z) \mathrm{d}z = \int\limits_0^{z_0} \frac{\delta R(\mu)}{\delta \kappa(z)} \kappa_m(z) \, \delta \ln f_m(z) \, \mathrm{d}z \tag{4.36}$$

Comparing this result with the general definition of the sensitivity to the continuous parameter, Eq. (2.6), we obtain the expression for the sensitivity with respect to mixing ratio in the form:

$$\frac{\delta R(\mu)}{\delta \ln f_m(z)} = \frac{\delta R(\mu)}{\delta \kappa(z)} \kappa_m(z) \tag{4.37}$$

Substituting Eq. (4.29), we obtain:

$$\frac{\delta R(\mu)}{\delta \kappa(z)} = -\kappa_m(z) \frac{1}{\mu} [r(z, \mu) - t(z, \mu) B(z)] \tag{4.38}$$

In the next chapter we will see more examples of relations between sensitivities of model input parameters and geophysical parameters.

References

The linear and non-linear demo models were developed by the author for the course on sensitivity analysis presented at the Jet Propulsion Laboratory a few years ago. The model of radiances of a non-scattering planetary atmosphere exists in different forms since middle of last century. Their sensitivity analysis in the form presented in this chapter was also developed for the above course.

Chapter 5
Sensitivity Analysis of Analytic Models: Linearization and Adjoint Approaches

Abstract In this chapter we apply the linearization and adjoint approaches of sensitivity analysis to three analytic models, for which we have obtained the solutions earlier, in Chap. 4, using the formalism of differential calculus (for discrete parameters) and variational calculus (for continuous parameters). We will reproduce these results using the approaches of sensitivity analysis described in general form in Chap. 3, which will provide a form of validation of these approaches.

Keywords Analytic models · Linearization approach · Adjoint approach

5.1 Linear Demo Model

5.1.1 Linearization Approach

Consider the simplest linear demo model with the forward problem Eq. (4.1) and observable Eq. (4.2). Assuming that the system Eq. (4.1) experiences a variation due to the variation of an arbitrary model input parameter, we have:

$$\begin{cases} \delta\left(\dfrac{\mathrm{d}X}{\mathrm{d}t} + a(t)X(t)\right) = 0 \\ \delta X(t_0) = \delta X_0 \end{cases} \qquad (5.1)$$

Rewriting Eq. (5.1) in the form

$$\begin{cases} \dfrac{\mathrm{d}\delta X}{\mathrm{d}t} + \delta a(t)X(t) + a(t)\delta X(t) = 0 \\ \delta X(t_0) = \delta X_0 \end{cases} \qquad (5.2)$$

and re-arranging the terms of the differential equation in Eq. (5.2), we obtain the corresponding linearized forward problem:

© The Author(s) 2015
E.A. Ustinov, *Sensitivity Analysis in Remote Sensing*,
SpringerBriefs in Earth Sciences, DOI 10.1007/978-3-319-15841-9_5

$$\begin{cases} \dfrac{d\delta X}{dt} + a(t)\delta X(t) = -X(t)\delta a(t) \\ \delta X(t_0) = \delta X_0 \end{cases} \tag{5.3}$$

The forward problem Eq. (4.1) has two model input parameters: one discrete parameter X_0 and one continuous parameter $a(t)$. We first consider linearization of the forward problem with respect to the discrete parameter X_0. The linearized problem Eq. (5.3) takes the form:

$$\begin{cases} \dfrac{d\delta_{X_0}X}{dt} + a(t)\delta_{X_0}X(t) = 0 \\ \delta_{X_0}X(t_0) = \delta X_0 \end{cases} \tag{5.4}$$

Its solution is obtained in the same way as the solution of the baseline forward problem Eq. (4.1):

$$\delta_{X_0}X(t) = \delta X_0 \exp[-A(t)] \tag{5.5}$$

where $A(t)$ is given by Eq. (4.4). Then we have:

$$\frac{\partial X(t)}{\partial X_0} = \exp[-A(t)] \tag{5.6}$$

and sensitivity to the discrete parameter X_0 is:

$$\frac{\partial R}{\partial X_0} = \frac{\partial X(t_1)}{\partial X_0} = \exp[-A(t_1)] \tag{5.7}$$

Comparing with expressions for the forward solution, Eq. (4.3), and for the observable, Eq. (4.2) we obtain:

$$\frac{\partial R}{\partial X_0} = \frac{X(t_1)}{X_0} = \frac{R}{X_0} \tag{5.8}$$

Thus, we obtained the result obtained earlier in Chap. 4, using the conventional approach of differential calculus.

Now we consider linearization of the forward problem Eq. (4.1) with respect to the continuous parameter $a(t)$. The linearized problem Eq. (5.3) takes the form:

$$\begin{cases} \dfrac{d\delta_a X}{dt} + a(t)\delta_a X(t) = -X(t)\delta a(t) \\ \delta_a X(t_0) = 0 \end{cases} \tag{5.9}$$

The differential equation of the linearized forward problem Eq. (5.9) has the form of an ODE of first order with a variable coefficient Eq. (4.15) and is solved using the same approach. Introducing an auxiliary function $M(t)$, which satisfies the condition $\dot{M} = a(t)M(t)$, we have:

$$M(t) = \exp[-A(t)] \tag{5.10}$$

and

$$\delta_a X(t) = \frac{\int_{t_0}^{t_1} M(t')[-X(t')\delta a(t')] \, dt' + C}{M(t)} \tag{5.11}$$

The initial-value condition of the linearized forward problem Eq. (5.9) yields $C = 0$, and substituting Eq. (5.10), we have

$$\delta_a X(t) = -\int_{t_0}^{t_1} \exp\{-[A(t') - A(t)]\}X(t')\delta a(t') \, dt' \tag{5.12}$$

Substituting the solution $X(t')$ Eq. (4.3) of the baseline forward problem Eq. (4.3), we have:

$$\delta_a X(t) = -X_0 \int_{t_0}^{t_1} \exp[-A(t')]\delta a(t') \, dt' \tag{5.13}$$

Comparing Eq. (5.13) with the definition of the variational derivative, Eq. (2.6) we obtain:

$$\frac{\delta X(t)}{\delta a(t')} = -X_0 \exp[-A(t')] \tag{5.14}$$

and

$$\frac{\delta R}{\delta a(t)} = \frac{\delta X(t_1)}{\delta a(t)} = -X_0 \exp[-A(t_1)] \tag{5.15}$$

Finally, comparing with expressions for the forward solution, Eq. (4.3), and for the observable, Eq. (4.2), we obtain:

$$\frac{\delta R}{\delta a(t)} = -R \tag{5.16}$$

Thus, we arrived at the result that was obtained earlier in Chap. 4, using the conventional approach of variational calculus.

5.1.2 Adjoint Approach

Now we consider the application of the adjoint approach to this linear demo model. To apply the Lagrange identity Eq. (3.20) and obtain the adjoint operator L^* of the corresponding adjoint problem, we need to explicitly build the corresponding linear operator L of the operator equation Eq. (3.18). Observing that, on the one hand, the equation of the linear forward problem Eq. (4.1) is acting on the whole interval $[t_0, t_1]$, and, on the other hand, the initial-value condition in Eq. (4.1) is acting only at the instant t_0, we can assemble the operator in the form:

$$L = \frac{d}{dt} + a(t) + \delta(t - t_0) \cdot 1 \tag{5.17}$$

Here the delta-function $\delta(t - t_0)$ ensures that the initial-value condition is applied to the solution $X(t)$ only at the instant t_0.

Assuming that $X(t)$ and $X^*(t)$ are two arbitrary functions specified on the interval $[t_0, t_1]$, we can write the left-hand term of the Lagrange identity Eq. (3.20) in the form:

$$
\begin{aligned}
(X^*, LX) &= \int_{t_0}^{t_1} X^*(t) \left[\frac{d}{dt} + a(t) + \delta(t - t_0) \right] X(t) \; dt \\
&= \int_{t_0}^{t_1} X^*(t) \frac{dX}{dt} \; dt + \int_{t_0}^{t_1} X^*(t) a(t) X(t) \; dt + X^*(t) X(t)|_{t_0}
\end{aligned}
\tag{5.18}
$$

Integrating the first term in Eq. (5.18) by parts, combining the resulting integral and off-integral terms, and cancelling out the off-integral terms at t_0 we obtain:

$$
\begin{aligned}
(X^*, LX) &= -\int_{t_0}^{t_1} \frac{dX^*}{dt} X(t) \; dt + \int_{t_0}^{t_1} a(t) X^*(t) X(t) \; dt + X^*(t) X(t)|_{t_1} \\
&= \int_{t_0}^{t_1} \left[-\frac{d}{dt} + a(t) + \delta(t - t_1) \right] X^*(t) X(t) \; dt = (L^* X^*, X)
\end{aligned}
\tag{5.19}
$$

where

$$L^* = -\frac{d}{dt} + a(t) + \delta(t - t_1) \cdot 1 \tag{5.20}$$

is the sought-for operator L^*, which is adjoint to the operator L, Eq. (5.17). The form of Eq. (5.20) implies that the adjoint problem has a final-value condition, rather than an initial-value condition as in the forward problem, Eq. (4.1). In order to formulate the adjoint problem in the form of a system of a differential equation and corresponding final-value condition, we need to formulate the right-hand term W of the adjoint problem Eq. (3.21). The expression for the observable, Eq. (4.2), can be rewritten in the form:

$$R = \int_{t_0}^{t_1} X(t)\delta(t - t_1) \, dt \qquad (5.21)$$

which means that the function $W(t)$ describing the observables procedure here has the form:

$$W(t) = 0 + \delta(t - t_1) \cdot 1 \qquad (5.22)$$

Now we can assemble the adjoint problem, corresponding to the forward problem Eq. (4.1) and the form of the observable Eq. (4.2) in the explicit form as

$$\begin{cases} -\dfrac{dX^*}{dt} + a(t)X^*(t) = 0 \\ X^*(t_1) = 1 \end{cases} \qquad (5.23)$$

The solution $X^*(t)$ of the adjoint problem Eq. (5.23) is obtained in the same manner as that of the baseline forward problem, Eq. (4.1), but the integration proceeds backwards from the final instant t_1 to the initial instant t_0. We have:

$$X^*(t) = \exp\left[-\int_{t}^{t_1} a(t') \, dt'\right] = \exp[A(t) - A(t_1)] \qquad (5.24)$$

or, using Eqs. (4.3) and (4.2):

$$X^*(t) = \frac{X(t_1)}{X(t)} = \frac{R}{X(t)} \qquad (5.25)$$

The right-hand terms of the forward problem Eq. (4.1) can be assembled in a fashion similar to Eq. (5.20):

$$S = 0 + \delta(t - t_0) \cdot X_0 \qquad (5.26)$$

Before computing the sensitivities, we can validate the adjoint solution Eq. (5.25) obtained above. Substituting Eq. (5.25) into the alternative expression for the observable, Eq. (3.22), we obtain an identity:

$$R = (X^*, S) = \int\limits_{t_0}^{t_1} \frac{R}{X(t)} \delta(t - t_0) X_0 \, dt = \frac{R}{X(t_0)} X_0 = R \qquad (5.27)$$

To obtain the sensitivity to the discrete parameter X_0 we use Eq. (3.25). From Eqs. (5.17) and (5.26) we have:

$$\frac{\partial L}{\partial X_0} = 0, \ \frac{\partial S}{\partial X_0} = 1$$

Substituting in Eq. (3.25) we have:

$$\frac{\partial R}{\partial X_0} = \left(X^*, \frac{\partial S}{\partial X_0} \right) = \int\limits_{t_0}^{t_1} X^*(t) \delta(t - t_0) \, dt = X^*(t_0) \qquad (5.28)$$

Substituting Eq. (5.25), we obtain the expression identical to Eqs. (4.6) and (5.8):

$$\frac{\partial R}{\partial X_0} = \frac{R}{X(t_0)} = \frac{R}{X_0} \qquad (5.29)$$

Thus, we obtained the result obtained earlier in Chap. 4 using the conventional approach of differential calculus, and above in this chapter using the linearization approach.

To obtain the sensitivity to the continuous parameter $a(t)$ we use Eq. (3.17). From Eqs. (5.17) and (5.26) we have:

$$\frac{\delta L}{\delta a(t)} = \delta(t' - t), \frac{\delta S}{\delta a(t)} = 0$$

Substituting in Eq. (3.26) we have:

$$\frac{\delta R}{\delta a(t)} = \left(X^*, -\frac{\delta L}{\delta a(t)} X \right) = -\int\limits_{t_0}^{t_1} X^*(t) \delta(t - t_0) X(t) \, dt = -X^*(t_0) X(t_0) \quad (5.30)$$

Substituting Eqs. (5.25) and (4.3), we obtain the expression identical to Eqs. (4.7) and (5.16):

$$\frac{\delta R}{\delta a(t)} = -\frac{R}{X(t_0)} X(t_0) = -R \qquad (5.31)$$

Thus, we obtained the result obtained earlier in Chap. 4 using the conventional approach of differential calculus, and above in this chapter using the linearization approach.

5.2 Non-linear Demo Model

5.2.1 Linearization Approach

Now we consider the simplest non-linear demo model with the forward problem Eq. (4.8) and observable Eq. (4.2). Assuming that the system Eq. (4.8) experiences a variation due to arbitrary variations of the model input parameters, we have:

$$\begin{cases} \delta\left(\dfrac{dX}{dt} + a(t)X^2(t)\right) = 0 \\ \delta X(t_0) = \delta X_0 \end{cases} \tag{5.32}$$

Rewriting Eq. (5.32) in the form

$$\begin{cases} \dfrac{d\delta X}{dt} + \delta a(t)X^2(t) + 2a(t)X(t)\delta X(t) = 0 \\ \delta X(t_0) = \delta X_0 \end{cases} \tag{5.33}$$

and re-arranging the terms of the differential equation in Eq. (5.33), we obtain the corresponding linearized forward problem:

$$\begin{cases} \dfrac{d\delta X}{dt} + 2a(t)X(t)\delta X(t) = -X^2(t)\delta a(t) \\ \delta X(t_0) = \delta X_0 \end{cases} \tag{5.34}$$

The forward problem Eq. (4.8) has two model input parameters: X_0 and $a(t)$. We first consider the linearization with respect to the discrete parameter X_0. The linearized problem Eq. (5.34) takes the form:

$$\begin{cases} \dfrac{d\delta_{X_0}X}{dt} + 2a(t)X(t)\delta_{X_0}X(t) = 0 \\ \delta_{X_0}X(t_0) = \delta X_0 \end{cases} \tag{5.35}$$

The solution of Eq. (5.35) is immediately obtained in the form:

$$\delta_{X_0}X(t) = \delta X_0 \exp\left[-2\int_{t_0}^{t} a(t')X(t')\, dt'\right] \tag{5.36}$$

Transforming the integral in Eq. (5.36) as

$$\int_{t_0}^{t} a(t')X(t') \, dt' = \int_{t_0}^{t} X(t') \, dA(t') \tag{5.37}$$

where $A(t)$ is given by Eq. (4.4), and substituting the solution $X(t)$ of the baseline forward problem Eq. (4.8) we can rewrite Eq. (5.37) in the form

$$\int_{t_0}^{t} a(t')X(t') \, dt' = \int_{t_0}^{t} \frac{dA(t')}{A(t') + X_0^{-1}} = \ln\left[A(t') + X_0^{-1}\right]\Big|_{t_0}^{t} \tag{5.38}$$

Comparing this result with Eq. (4.9), we have:

$$\int_{t_0}^{t} a(t')X(t') \, dt' = \ln\left(\frac{X_0}{X(t)}\right) \tag{5.39}$$

Substituting Eq. (5.39) in Eq. (5.36), and observing that $\delta_{X_0}X(t_0) = \delta X_0$ we obtain

$$\delta_{X_0}X(t) = \delta X_0 \left(\frac{X(t)}{X_0}\right)^2 \tag{5.40}$$

Then

$$\frac{\partial R}{\partial X_0} = \frac{\partial X(t_1)}{\partial X_0} = \left(\frac{X(t_1)}{X_0}\right)^2 \tag{5.41}$$

Comparing with the expression for the observable, Eq. (4.2) we have:

$$\frac{\partial R}{\partial X_0} = \left(\frac{R}{X_0}\right)^2 \tag{5.42}$$

Thus, we obtained the result obtained earlier in Chap. 4 using the conventional approach of differential calculus.

Now we consider linearization of the forward problem Eq. (4.8) with respect to the continuous parameter $a(t)$. The linearized problem Eq. (5.34) takes the form:

$$\begin{cases} \dfrac{d\delta_a X}{dt} + 2a(t)X(t)\delta_a X(t) = -X^2(t)\delta a(t) \\ \delta_a X(t_0) = 0 \end{cases} \tag{5.43}$$

The differential equation of this linearized forward problem has the form of an ODE of first order with a variable coefficient Eq. (4.15). Introducing an auxiliary function

$M(t)$, which satisfies the condition $\dot{M} = 2a(t)X(t)M(t)$, we use the standard method of solution presented in Sect. 4.3. Substituting $p(t) = 2a(t)X(t)M(t)$ and applying the log function to both sides of the resulting equation we have:

$$\ln M(t)\big|_{t_0}^{t} = -2 \int\limits_{t_0}^{t} a(t')X(t')\ dt' \tag{5.44}$$

Using Eq. (5.39) and discarding the non-essential factors $M(t_0)$ and X_0 we obtain:

$$M(t) = \frac{1}{X^2(t)} \tag{5.45}$$

Now, writing Eq. (4.17) for the variation $\delta_a X(t)$, and substituting $q(t) = -X^2(t)\delta a(t)$, we have:

$$\delta_a X(t) = \frac{\int_{t_0}^{t_1} M(t')[-X^2(t)\delta a(t')]\ dt' + C}{M(t)} \tag{5.46}$$

The initial-value condition of the linearized forward problem Eq. (5.43) yields $C = 0$, and substituting Eq. (5.45) into Eq. (5.46), we have

$$\delta_a X(t) = -X^2(t) \int\limits_{t_0}^{t_1} X^{-2}(t')X^2(t')\delta a(t')\ dt' = \int\limits_{t_0}^{t_1} \left[-X^2(t)\right]\delta a(t')\ dt' \tag{5.47}$$

Comparing Eq. (5.47) with the definition of the variational derivative, Eq. (4.6) we obtain the expression, which is identical to Eq. (4.12):

$$\frac{\delta R}{\delta a(t)} = \frac{\delta X(t_1)}{\delta a(t)} = -X^2(t_1) = -R^2 \tag{5.48}$$

Thus, we arrived at the result, which was obtained earlier in Chap. 4 using the conventional approach of variational calculus.

5.2.2 Adjoint Approach

Now we consider the application of the adjoint approach to this non-linear demo model. We assemble the operator of the corresponding linearized forward problem Eq. (5.34) in the form:

$$L = \frac{\mathrm{d}}{\mathrm{d}t} + 2a(t)X(t) + \delta(t - t_0) \cdot 1 \tag{5.49}$$

Assuming that $X(t)$ and $X^*(t)$ are two arbitrary functions specified on the interval $[t_0, t_1]$, as in the derivation of the adjoint operator L^* for the linear demo model above, after similar manipulations we obtain:

$$(X^*, LX) = \int\limits_{t_0}^{t_1} \left[-\frac{\mathrm{d}}{\mathrm{d}t} + 2a(t)X(t) + \delta(t - t_1) \right] X^*(t) X(t) \ \mathrm{d}t = (L^*X^*, X) \tag{5.50}$$

where

$$L^* = -\frac{\mathrm{d}}{\mathrm{d}t} + 2a(t)X(t) + \delta(t - t_1) \cdot 1 \tag{5.51}$$

is the sought-for operator L^*, which is adjoint to the operator L, Eq. (5.49). The expression for the observable R, Eq. (4.2), is presented in the form of Eq. (5.22). Now we can assemble the adjoint problem, corresponding to the linearized forward problem Eq. (5.34) in explicit form as

$$\begin{cases} -\dfrac{\mathrm{d}X^*}{\mathrm{d}t} + 2a(t)X(t)X^*(t) = 0 \\ X^*(t_1) = 1 \end{cases} \tag{5.52}$$

The solution $X^*(t)$ of the adjoint problem Eq. (5.52) is obtained in the same manner as that of the linearized forward problem, Eq. (5.34), but the integration, again, proceeds backwards from the final instant t_1 to the initial instant t_0. We have [cf. Eq. (5.36)]:

$$X^*(t) = \exp\left[-2 \int\limits_{t}^{t_1} a(t')X(t') \ \mathrm{d}t' \right] = \exp[A(t) - A(t_1)] \tag{5.53}$$

or, using Eqs. (4.3) and (4.2):

$$X^*(t) = \left(\frac{X(t_1)}{X(t)} \right)^2 = \left(\frac{R}{X(t)} \right)^2 \tag{5.54}$$

The right-hand terms of the linearized forward problem Eq. (5.34) can be assembled in a fashion similar to Eq. (5.49):

$$\delta S - \delta N[X] = -X^2 \delta a + \delta(t - t_0)\delta X_0 \tag{5.55}$$

To obtain the sensitivity to the discrete parameter X_0 we use the definition Eq. (3.25). From Eq. (5.55) we have:

$$\frac{\partial S}{\partial X_0} - \frac{\partial N}{\partial X_0} X = \delta(t - t_0)$$

Substituting in Eq. (3.25) we have:

$$\frac{\partial R}{\partial X_0} = \int_{t_0}^{t_1} X^*(t)\delta(t - t_0)\ \mathrm{d}t = X^*(t_0) \tag{5.56}$$

Substituting Eq. (5.54), we obtain the expression identical to Eqs. (4.11) and (5.42):

$$\frac{\partial R}{\partial X_0} = \left(\frac{R}{X(t_0)}\right)^2 = \left(\frac{R}{X_0}\right)^2 \tag{5.57}$$

Thus, we obtained the result obtained earlier in Chap. 4 using the conventional approach of differential calculus, and above in this chapter using the linearization approach of sensitivity analysis.

To obtain the sensitivity to the continuous parameter $a(t)$ we use the definition Eq. (3.26). From Eq. (5.55) we have:

$$\frac{\delta S}{\delta a(t')} - \frac{\delta N}{\delta a(t')} X = -\delta(t' - t)X^2(t')$$

Substituting in Eq. (3.26), we have:

$$\frac{\delta R}{\delta a(t)} = -\int_{t_0}^{t_1} X^*(t')\delta(t' - t)X^2(t')\ \mathrm{d}t' = -X^*(t)X^2(t) \tag{5.58}$$

Substituting Eq. (5.54) and cancelling out the equal factors $X^2(t)$ in the numerator and denominator of the resulting expression, we obtain the expression identical to Eqs. (4.12) and (5.48):

$$\frac{\delta R}{\delta a(t)} = -R^2 \tag{5.59}$$

Thus, we obtained the result obtained earlier in Chap. 4 using the conventional approach of differential calculus, and above in this chapter using the linearization approach of sensitivity analysis.

5.3 Model of Radiances of a Non-scattering Planetary Atmosphere

5.3.1 Linearization Approach

Now we consider the model with the forward problem Eq. (4.13) and observable Eq. (4.14). Assuming that the system Eq. (4.13) experiences a variation due to arbitrary variations of the model input parameters, we have:

$$
\begin{cases}
\delta\left(u\dfrac{dI}{dz} + \kappa(z)I(z,u)\right) = \delta(\kappa(z)B(z)) \\
\delta I(0,u) = 0, \quad u > 0 \\
\delta I(z_0,u) = \delta(\varepsilon B_s), \quad u < 0
\end{cases}
\tag{5.60}
$$

Rewriting Eq. (5.60) in the form

$$
\begin{cases}
u\dfrac{d\delta I}{dz} + \delta\kappa(z)I(z,u) + \kappa(z)\delta I(z,u) = \delta\kappa(z)B(z) + \kappa(z)\delta B(z) \\
\delta I(0,u) = 0, \quad u > 0 \\
\delta I(z_0,u) = \varepsilon\delta B_s + B_s\delta\varepsilon, \quad u < 0
\end{cases}
\tag{5.61}
$$

and re-arranging the terms of the differential equation in Eq. (5.61), we obtain the corresponding linearized forward problem:

$$
\begin{cases}
u\dfrac{d\delta I}{dz} + \kappa(z)\delta I(z,u) = \kappa(z)\delta B(z) + [B(z) - I(z,u)]\delta\kappa(z) \\
\delta I(0,u) = 0, \quad u > 0 \\
\delta I(z_0,u) = \varepsilon\delta B_s + B_s\delta\varepsilon, \quad u < 0
\end{cases}
\tag{5.62}
$$

The forward problem Eq. (4.13) has four model input parameters: two discrete parameters: B_s and ε, and two continuous parameters: $B(z)$ and $\kappa(z)$. We first consider the linearization with respect to the discrete parameters B_s and ε. For B_s, the linearized problem Eq. (5.62) takes the form:

$$
\begin{cases}
u\dfrac{d\delta_{B_s}I}{dz} + \kappa(z)\delta_{B_s}I(z,u) = 0 \\
\delta_{B_s}I(0,u) = 0, \quad u > 0 \\
\delta_{B_s}I(z_0,u) = \varepsilon\delta B_s, \quad u < 0
\end{cases}
\tag{5.63}
$$

The solution of the problem Eq. (5.63) for $u < 0$ is immediately obtained in the form:

$$\delta_{B_s} I(z, u) = \exp\left(\frac{1}{u}\int_z^{z_0} \kappa(z')dz'\right)\varepsilon\delta B_s \tag{5.64}$$

Correspondingly,

$$\delta_{B_s} R(\mu) = \exp\left(-\frac{1}{\mu}\int_0^{z_0} \kappa(z')dz'\right)\varepsilon\delta B_s = t(z_0, \mu)\varepsilon\delta B_s \tag{5.65}$$

and we obtain

$$\frac{\partial R}{\partial B_s} = t(z_0, \mu)\varepsilon \tag{5.66}$$

in accordance with Eq. (4.21).
   For ε, the linearized problem Eq. (5.62) takes the form:

$$\begin{cases} u\dfrac{d\delta_\varepsilon I}{dz} + \kappa(z)\delta_\varepsilon I(z, u) = 0 \\ \delta_\varepsilon I(0, u) = 0, \quad u > 0 \\ \delta_\varepsilon I(z_0, u) = B_s\delta\varepsilon, \quad u < 0 \end{cases} \tag{5.67}$$

The solution of Eq. (5.67) for $u < 0$ is immediately obtained in the form:

$$\delta_\varepsilon I(z, u) = \exp\left(\frac{1}{u}\int_z^{z_0} \kappa(z')dz'\right)B_s\delta\varepsilon \tag{5.68}$$

Correspondingly,

$$\delta_\varepsilon R(\mu) = \exp\left(-\frac{1}{\mu}\int_0^{z_0} \kappa(z')dz'\right)B_s\delta\varepsilon = t(z_0, \mu)B_s\delta\varepsilon \tag{5.69}$$

and we obtain

$$\frac{\partial R}{\partial \varepsilon} = t(z_0, \mu)B_s \tag{5.70}$$

in accordance with Eq. (4.22).

Now we consider the linearization of the forward problem Eq. (4.8) with respect to the continuous parameters $B(z)$ and $\kappa(z)$. For $B(z)$, the linearized problem Eq. (5.62) takes the form:

$$
\begin{cases}
u\dfrac{d\delta_B I}{dz} + \kappa(z)\delta_B I(z, u) = \kappa(z)\delta B(z) \\
\delta_P I(0, u) = 0, \quad u > 0 \\
\delta_P I(z_0, u) = 0, \quad u < 0
\end{cases}
\tag{5.71}
$$

The solution of Eq. (5.71) for $u < 0$ is obtained in the form:

$$
\delta_B I(z, u) = -\frac{1}{u}\int\limits_{z}^{z_0} \delta B(z') \exp\left(\frac{1}{u}\int\limits_{z'}^{z}\kappa(z'')dz''\right)\kappa(z')dz'
\tag{5.72}
$$

Correspondingly,

$$
\delta_B R(\mu) = \frac{1}{\mu}\int\limits_{z}^{z_0}\delta B(z)\exp\left(-\frac{1}{\mu}\int\limits_{0}^{z}\kappa(z')dz'\right)\kappa(z)dz = \frac{1}{\mu}\int\limits_{z}^{z_0}t(z_0,\mu)\kappa(z)\delta B(z)dz
\tag{5.73}
$$

and we obtain

$$
\frac{\delta R}{\delta B(z)} = \frac{1}{\mu}\kappa(z)t(z_0,\mu)
\tag{5.74}
$$

in accordance with Eq. (4.23).

For $\kappa(z)$, the linearized problem Eq. (5.62) takes the form:

$$
\begin{cases}
u\dfrac{d\delta_\kappa I}{dz} + \kappa(z)\delta_\kappa I(z, u) = [B(z) - I(z, u)]\delta\kappa(z) \\
\delta_\kappa I(0, u) = 0, \quad u > 0 \\
\delta_\kappa I(z_0, u) = 0, \quad u < 0
\end{cases}
\tag{5.75}
$$

The solution of Eq. (5.75) for $u < 0$ is obtained in the form:

$$
\delta_\kappa I(z, u) = -\frac{1}{u}\int\limits_{z}^{z_0}[B(z') - I(z', u)]\delta\kappa(z')\exp\left(\frac{1}{u}\int\limits_{z'}^{z}\kappa(z'')dz''\right)\kappa(z')dz'
\tag{5.76}
$$

Correspondingly,

$$
\begin{aligned}
\delta_\kappa R(\mu) &= \frac{1}{\mu} \int_0^{z_0} [B(z) - I(z, -\mu)] \delta\kappa(z) \exp\left(-\frac{1}{\mu} \int_0^z \kappa(z')dz' \right) \kappa(z)dz \\
&= \frac{1}{\mu} \int_0^{z_0} \delta\kappa(z) t(z, \mu)(B(z) - I(z, -\mu)) \delta\kappa(z)dz
\end{aligned}
\tag{5.77}
$$

and we obtain

$$
\frac{\delta R(\mu)}{\delta\kappa(z)} = \frac{1}{\mu} t(z, \mu)[B(z) - I(z, -\mu)]
\tag{5.78}
$$

To reduce Eq. (5.78) to the form of Eq. (4.29) we write Eq. (4.18) for $\mu = -u$:

$$
\begin{aligned}
I(z, -\mu) &= R(\mu) \\
&= \frac{1}{\mu} \int_z^{z_0} B(z') \exp\left(-\frac{1}{\mu} \int_z^{z'} \kappa(z'')dz'' \right) \kappa(z')dz' + \varepsilon B_s \exp\left(-\frac{1}{\mu} \int_z^{z_0} \kappa(z')dz' \right)
\end{aligned}
\tag{5.79}
$$

Substituting Eq. (5.79) in Eq. (5.78), opening the parentheses in the resulting expression and evaluating the term $t(z, \mu)I(z, -\mu)$ we have:

$$
\begin{aligned}
t(z, \mu)I(z, -\mu) &= \exp\left(-\frac{1}{\mu} \int_0^z \kappa(z')dz' \right) \\
&\quad \times \left[\frac{1}{\mu} \int_z^{z_0} B(z') \exp\left(-\frac{1}{\mu} \int_z^{z'} \kappa(z'')dz'' \right) \kappa(z')dz' \right. \\
&\qquad \left. + \varepsilon B_s \exp\left(-\frac{1}{\mu} \int_z^{z_0} \kappa(z')dz' \right) \right] \\
&= \frac{1}{\mu} \int_z^{z_0} B(z') \exp\left(-\frac{1}{\mu} \int_0^{z'} \kappa(z'')dz'' \right) \kappa(z')dz' \\
&\quad + \varepsilon B_s \exp\left(-\frac{1}{\mu} \int_0^{z_0} \kappa(z')dz' \right)
\end{aligned}
\tag{5.80}
$$

Using the definitions of the atmospheric transmittance, Eq. (4.20), and of the upwelling radiance, Eq. (4.30), we have:

$$
\begin{aligned}
t(z,\mu)I(z,-\mu) &= \frac{1}{\mu}\int_z^{z_0} B(z')t(z,\mu)\kappa(z')\mathrm{d}z' + \varepsilon B_s t(z_0,\mu) \\
&= \int_z^{z_0} B(z')\mathrm{d}t(z,\mu) + \varepsilon B_s t(z_0,\mu) = r(z,\mu)
\end{aligned}
\tag{5.81}
$$

Then, Eq. (5.78) can be rewritten in the form:

$$
\frac{\delta R}{\delta \kappa(z)} = -\frac{1}{\mu}[r(z,\mu) - t(z,\mu)B(z)]
\tag{5.82}
$$

in accordance with Eq. (4.29).

5.3.2 Adjoint Approach

To formulate the corresponding adjoint problem, we rewrite the forward problem, Eq. (4.13), in a general form

$$
L_e I - S_e = 0
\tag{5.83}
$$

$$
L_t I - S_t = 0, \; z = 0, \quad u > 0
\tag{5.84}
$$

$$
L_b I - S_b = 0, \; z = z_0, \quad u < 0
\tag{5.85}
$$

where the subscripts 'e', 't' and 'b' stand for 'equation', upper ('top') and lower ('bottom') boundary conditions respectively. Operators L_e, L_t, and L_b have the form:

$$
L_e I = u\frac{\mathrm{d}I}{\mathrm{d}z} + \kappa(z)I(z,u)
\tag{5.86}
$$

$$
L_t I = I(0,u), \quad u > 0
\tag{5.87}
$$

$$
L_b I = I(z_0,u), \quad u < 0
\tag{5.88}
$$

The right-hand terms S_e, S_t, and S_b have the form:

$$
S_e = \kappa(z)B(z)
\tag{5.89}
$$

$$S_t = 0 \tag{5.90}$$

$$S_b = \varepsilon B_s \tag{5.91}$$

The operator L_e contains the linear operations on $I(z, u)$ over the whole domain of arguments $z \in [0, z_0]$ and $u \in [-1, 1]$. On the other hand, the operators L_t and L_b are limited in action with respect to both arguments z and u. The operator L_t is an identity operator limited in its action to the upper boundary and upper hemisphere. Similarly, the action of the operator L_b is limited to the lower boundary and lower hemisphere. To enforce these restrictions, we multiply Eqs. (5.84) and (5.85) by appropriate weighting factors containing Dirac delta functions over z, and Heaviside θ-functions over u:

$$\delta(z)\theta(u)(L_tI - S_t) = 0 \tag{5.92}$$

$$\delta(z - z_0)\theta(-u)(L_bI - S_b) = 0 \tag{5.93}$$

To proceed, we observe that Eqs. (5.92) and (5.93) remain equivalent to the boundary conditions, Eqs. (5.84) and (5.85), if they are multiplied by arbitrary functions of z and/or u that are non-zero everywhere, where the above weighting factors are non-zero. To facilitate the application of Lagrange identity, Eq. (3.20), we multiply Eqs. (5.92) and (5.93) by factors u and $-u$ respectively. We have:

$$\delta(z)\lambda(u)(L_tI - S_t) = 0 \tag{5.94}$$

$$\delta(z - z_0)\lambda(-u)(L_bI - S_b) = 0 \tag{5.95}$$

where the λ-function introduced in (Ustinov 2001)

$$\lambda(u) = u\theta(u) \tag{5.96}$$

is the primitive function (antiderivative) of $\theta(u)$:

$$\frac{d\lambda}{du} = \theta(u) \tag{5.97}$$

Adding Eqs. (5.83), (5.94) and (5.95) together, and moving the terms with S_e, S_t, and S_b into the right-hand side, we obtain the linear operator equation in the form of Eq. (3.18), where the linear operator L contains all operations on I:

$$LI = L_eI + \delta(z)\lambda(u)L_tI + \delta(z - z_0)\lambda(-u)L_bI \tag{5.98}$$

and the right-hand term

$$S = S_e + \delta(z)\lambda(u)S_t + \delta(z - z_0)\lambda(-u)S_b \qquad (5.99)$$

is a scalar function of arguments z and u. Substituting Eqs. (5.86)–(5.88) in Eq. (5.98) and Eqs. (5.89)–(5.91) in Eq. (5.99), we have:

$$LI = u\frac{dI}{dz} + \kappa(z)I(z,u) + \delta(z)\lambda(u)I(z,u) + \delta(z - z_0)\lambda(-u)I(z,u) \qquad (5.100)$$

$$S = \kappa(z)B(z) + \delta(z)\lambda(u) \cdot 0 + \delta(z - z_0)\lambda(-u)\varepsilon B_s \qquad (5.101)$$

The observable $R(\mu)$ is represented in the form of the inner product

$$R(\mu) = (W, I) = \int\limits_0^{z_0} dz \int\limits_{-1}^{1} du W(z, u)I(z, u) \qquad (5.102)$$

where the observables weighting function

$$W(z, u) = \delta(z)\theta(u + \mu) \qquad (5.103)$$

corresponds to the measurements at the top of the atmosphere $z = 0$, in the direction $u = -\mu$ opposite to the viewing direction μ.

To obtain the operator L^* of the corresponding adjoint problem we apply the Lagrange identity, written here in the form:

$$(I^*, LI) = (L^*I^*, I) \qquad (5.104)$$

Performing integrations of terms containing the δ-function, we have:

$$(I^*, LI) = \int\limits_{-1}^{1} du \int\limits_0^{z_0} dz I^* \left(u\frac{d}{dz} + \kappa \right)I + \int\limits_0^{1} du \cdot uI^*I \bigg|_{z=0} + \int\limits_{-1}^{0} du \cdot uI^*I \bigg|_{z=z_0}$$

$$(5.105)$$

Integrating by parts

$$\int\limits_0^{z_0} I^*\frac{dI}{dz}dz = I^*I|_{z=0}^{z=z_0} - \int\limits_0^{z_0} \frac{dI^*}{dz}Idz$$

we can rewrite Eq. (5.105) in the form:

$$
(I^*, LI) = \int_{-1}^{1} du \int_{0}^{z_0} dz \left(-u\frac{d}{dz} + \kappa\right) I^* I + \int_{-1}^{1} du \cdot u I^* I \Big|_{z=0}^{z=z_0}
$$

$$
+ \int_{0}^{1} du \cdot u I^* I \Big|_{z=0} + \int_{-1}^{0} du \cdot u I^* I \Big|_{z=z_0}
$$

$$
= \int_{-1}^{1} du \int_{0}^{z_0} dz \left(-u\frac{d}{dz} + \kappa\right) I^* I + \int_{-1}^{0} du \cdot u I^* I \Big|_{z=0} + \int_{0}^{1} du \cdot u I^* I \Big|_{z=z_0} \tag{5.106}
$$

The resulting expression in Eq. (5.106) is represented in the desired form of the inner product (L^*I^*, I) on the right-hand side of the Lagrange identity, Eq. (3.20), where the operator L^* of the corresponding adjoint problem has the form:

$$
L^*I^* = L_e^*I^* + \delta(z)\lambda(-u)L_t^*I^* + \delta(z - z_0)\lambda(u)L_b^*I^* \tag{5.107}
$$

Here, the operator

$$
L_e^*I^* = -u\frac{dI^*}{dz} + \kappa(z)I^*(z, u) \tag{5.108}
$$

contains the linear operations on I^* in the whole domain of $z \in [0, z_0]$ and $u \in [-1, 1]$, operators L_t^* and L_b^* are identity operators whose action is limited to the top boundary and lower hemisphere, and bottom boundary and upper hemisphere correspondingly.

The right-hand term W of the adjoint problem Eq. (3.21) is split here into its components corresponding to the differential equation and boundary conditions in the same fashion as the adjoint operator L^*, Eq. (5.107):

$$
W = W_e + \delta(z)\lambda(-u)W_t + \delta(z - z_0)\lambda(u)W_b \tag{5.109}
$$

Comparing Eq. (5.109) with the definition of W, Eq. (5.103), we have

$$
\delta(z)\delta(u + \mu) = W_e + \delta(z)\lambda(-u)W_t + \delta(z - z_0)\lambda(u)W_b \tag{5.110}
$$

whence for the components of the equation and bottom boundary condition we have:

$$
W_e = 0, \quad W_b = 0 \tag{5.111}
$$

and the only non-zero component is that corresponding to the top boundary condition. Cancelling out the factors $\delta(z)$ on both sides of the resulting equality, and using the definition of the λ-function, Eq. (5.96), we have:

$$\delta(u + \mu) = (-u)\theta(-u)W_t \tag{5.112}$$

Observing that the Heaviside function $\theta(x) = 1$ for $x > 0$ we have:

$$W = W_t = \frac{\delta(u + \mu)}{(-u)\theta(-u)} = \frac{\delta(u + \mu)}{(-u)} = \frac{1}{\mu}\delta(u + \mu) \tag{5.113}$$

Thus, the adjoint RT problem corresponding to the forward problem Eq. (4.13) and the observables weighting function, Eq. (5.113), has the form:

$$\begin{cases} -u\dfrac{dI^*}{dz} + \kappa(z)I^*(z, u) = 0 \\[2mm] I^*(0, u) = \dfrac{1}{\mu}\delta(u + \mu), \quad u < 0 \\[2mm] I^*(z_0, u) = 0, \quad u > 0 \end{cases} \tag{5.114}$$

Its solution $I^*(z, u)$ for $u < 0$ has the form:

$$\begin{aligned} I^*(z, u) &= \frac{1}{\mu}\delta(u + \mu)\exp\left(-\frac{1}{-u}\int_0^z \kappa(z')dz'\right) \\[3mm] &= \frac{1}{\mu}\delta(u + \mu)\exp\left(-\frac{1}{\mu}\int_0^z \kappa(z')dz'\right) = \frac{1}{\mu}\delta(u + \mu)t(z, \mu) \end{aligned} \tag{5.115}$$

To verify the obtained adjoint solution we compute the observable $R(\mu)$ using the alternative expression, Eq. (3.22). We have:

$$\begin{aligned} R(\mu) = (I^*, S) &= \int_0^{z_0} dz \int_{-1}^1 du I^*(z, u)S(z, u) \\[3mm] &= \frac{1}{\mu}\int_0^{z_0} t(z, \mu)S(z, -\mu)\, dz \\[3mm] &= \frac{1}{\mu}\int_0^{z_0} t(z, \mu)[\kappa(z)B(z) + \delta(z - z_0)\mu\varepsilon B_s]\, dz \\[3mm] &= \frac{1}{\mu}\int_0^{z_0} t(z, \mu)\kappa(z)B(z)\, dz + t(z_0, \mu)\varepsilon B_s \end{aligned} \tag{5.116}$$

The obtained expression is identical to Eq. (4.19).

To obtain the sensitivities to discrete parameters B_s and ε we use Eq. (3.25). For the surface Planck function B_s, from Eqs. (5.100) and (5.101), we have:

$$\frac{\partial L}{\partial B_s} = 0; \quad \frac{\partial S}{\partial B_s} = \delta(z - z_0)(-u)\theta(-u)\varepsilon$$

Substituting in Eq. (3.25), we have:

$$\frac{\partial R(\mu)}{\partial B_s} = \int\limits_0^{z_0} dz \int\limits_{-1}^{1} du I^*(z, u)\delta(z - z_0)(-u)\theta(-u)\varepsilon$$

$$= \frac{1}{\mu} t(z_0, \mu)\mu\varepsilon = t(z_0, \mu)\varepsilon \tag{5.117}$$

in accordance with Eqs. (4.21) and (5.66).

Similarly, for the surface emissivity ε from Eqs. (5.100) and (5.101) we have:

$$\frac{\partial L}{\partial \varepsilon} = 0; \quad \frac{\partial S}{\partial \varepsilon} = \delta(z - z_0)(-u)\theta(-u)B_s$$

Substituting in Eq. (3.25), we obtain:

$$\frac{\partial R(\mu)}{\partial \varepsilon} = \int\limits_0^{z_0} dz \int\limits_{-1}^{1} du I^*(z, u)\delta(z - z_0)(-u)\theta(-u)B_s = t(z_0, \mu)B_s \tag{5.118}$$

in accordance with Eqs. (4.22) and (5.70).

To obtain the sensitivities to continuous parameters $B(z)$ and $\kappa(z)$ we use Eq. (3.26). For the atmospheric Planck function $B(z)$, from Eq. (5.100) it follows that $\delta_B LI = 0$, and from Eq. (5.101) it follows that $\delta_B S = \kappa\delta B$. Therefore we have:

$$\frac{\delta L}{\delta B(z')}I = 0, \quad \frac{\delta S(z)}{\delta B(z')} = \delta(z' - z)\kappa(z) \tag{5.119}$$

Substituting Eqs. (5.115) and (5.119) in Eq. (3.26), we obtain:

$$\frac{\delta R(\mu)}{\delta B(z)} = \int\limits_0^{z_0} dz \int\limits_{-1}^{1} du I^*(z', u)\frac{\delta S(z)}{\delta B(z')}$$

$$= \int\limits_0^{z_0} dz'\delta(z' - z)\kappa(z) \int\limits_{-1}^{1} du \frac{1}{\mu}\delta(u + \mu)t(z, \mu) \tag{5.120}$$

$$= \frac{1}{\mu}\kappa(z)t(z, \mu)$$

in accordance with Eqs. (4.23) and (5.74).

For the absorption coefficient $\kappa(z)$ from Eq. (5.100) it follows that $\delta_\kappa LI = I\delta\kappa$, and from Eq. (5.101) it follows that $\delta_\kappa S = B\delta\kappa$. Therefore we have:

$$\frac{\delta L}{\delta \kappa(z')}I = \delta(z'-z)I(z,u), \quad \frac{\delta S(z)}{\delta \kappa(z')} = \delta(z'-z)B(z) \qquad (5.121)$$

Substituting Eqs. (5.115) and (5.121) in Eq. (3.26), we obtain:

$$
\begin{aligned}
\frac{\delta R(\mu)}{\delta \kappa(z)} &= \int\limits_0^{z_0} dz \int\limits_{-1}^1 du I^*(z',u)\left(\frac{\delta S(z)}{\delta \kappa(z')} - \frac{\delta L}{\delta \kappa(z')}I(z',u)\right) \\
&= \int\limits_0^{z_0} dz'\,\delta(z'-z)\int\limits_{-1}^1 du\,\frac{1}{\mu}\delta(u+\mu)t(z,\mu)[B(z)-I(z,u)] \\
&= \frac{1}{\mu}t(z,\mu)[B(z)-I(z,-\mu)]
\end{aligned}
\qquad (5.122)
$$

in accordance with Eq. (5.78) converted to Eqs. (4.29) and (5.82).

5.3.3 Summary

In this chapter we have applied two approaches of sensitivity analysis to three analytical models of which the first two are just simple demo models, but the third model has a definite practical importance. We have demonstrated that these approaches produce correct results by reproducing the analytic expressions for sensitivities to discrete and continuous parameters of these models, obtained earlier in Chap. 4 by direct application of differential and variational calculus to analytic expressions for observables of these models.

References

Sensitivity analysis of the linear and non-linear demo models using the linearization and adjoint approaches in the form presented in this chapter was developed by the author for the course on sensitivity analysis mentioned in the previous chapter. The concept of assembling the linear operator and the right-hand term of the operator forward problem in the form of weighted sums of the operators and right-hand terms of the differential equation(s) and initial and/or boundary conditions was developed in the author's paper:

Ustinov EA (2001) Adjoint sensitivity analysis of radiative transfer equation: temperature and gas mixing ratio weighting functions for remote sensing of scattering atmospheres in thermal IR. J Quant Spectr Rad Transf 68:195–211

Chapter 6
Sensitivity Analysis of Numerical Models

Abstract In the previous chapter, the linearization and adjoint approaches were applied to analytic models, and the obtained results matched those obtained in Chap. 4 using methods of differential calculus and variational calculus. This provided, albeit not rigorous but hopefully still convincing validation of general formulations of the linearization and adjoint approaches formulated in Chap. 3. In this chapter, the linearization and adjoint approaches of sensitivity analysis will be applied to selected numerical models for which analytic solutions are not available.

Keywords Numerical models · Linearization approach · Adjoint approach · Adjoint propagator

6.1 Model of Radiances of a Scattering Planetary Atmosphere

6.1.1 Baseline Forward Problem and Observables

The forward problem in this Section differs from that of the non-scattering model considered in Chaps. 4 and 5 by an additional term in the equation of radiative transfer due to atmospheric scattering. To add more generality, a term accounting for scattering on the underlying surface is included into the lower boundary condition. The corresponding forward problem has the form:

$$
\begin{cases}
u\dfrac{dI}{dz} + \alpha(z)\left(I(z,u) - \dfrac{1}{2}\displaystyle\int_{-1}^{1} p(z;u,u')I(z,u')du'\right) = \alpha(z)[1-\omega_0(z)]B(z) \\[4mm]
I(0,u) = 0, \quad \text{for } u > 0 \\[3mm]
I(z_0,u) - 2A\displaystyle\int_{0}^{1} I(z_0,u')u'du' = (1-A)B_s, \quad \text{for } u < 0
\end{cases}
\tag{6.1}
$$

© The Author(s) 2015
E.A. Ustinov, *Sensitivity Analysis in Remote Sensing*,
SpringerBriefs in Earth Sciences, DOI 10.1007/978-3-319-15841-9_6

Here, $\alpha(z)$ is the atmospheric extinction coefficient, $p(z; u, u')$ is the phase function of atmospheric scattering, and $\omega_0(z)$ is the single scattering albedo, which is the zero-th expansion coefficient of the phase function over Legendre coefficients

$$p(z; u, u') = \sum_{l=0}^{N} \omega_l(z) P_l(u) P_l(u') \tag{6.2}$$

In the lower boundary condition, A is the Lambertian surface albedo. The remaining entries in Eq. (6.1) are the same as in Eq. (4.13). Note that the right-hand term of the equation in the system Eq. (6.1) is identical to that of the non-scattering case, Eq. (4.13), because, by definition

$$\alpha(1 - \omega_0) = \kappa \tag{6.3}$$

Similarly, in the lower boundary condition of Eq. (6.1)

$$1 - A = \varepsilon \tag{6.4}$$

There is a wide variety of methods of solution of problems of radiative transfer. The majority of them represent the continuous angular dependence of the solution $I(z, u)$ by a finite-dimensional vector function of z. For example, the method of discrete ordinates represents the solution $I(z, u)$ by the vector $I(z)$ with components $I_j(z) = I(z, u_j)$, where $\{u_j\}$ is a set of nodes of an appropriate Gaussian quadrature. Once the solution is obtained in its finite-dimensional form, it can be computed for any angle of propagation $\cos^{-1} u$ using integration of the atmospheric source function, which for the equation of the system Eq. (6.1) has the form:

$$B(z, u) = \frac{1}{2} \int_{-1}^{1} p(z; u, u') I(z, u') du' + [1 - \omega_0(z)] B(z) \tag{6.5}$$

and is complemented by the corresponding surface source function derived from the lower boundary condition of Eq. (6.1):

$$B_s = 2A \int_{0}^{1} I(z_0, u') u' du' + (1 - A) B_s \tag{6.6}$$

Using Eqs. (6.5) and (6.6), we can rewrite the forward problem Eq. (6.1) in the form:

$$\begin{cases} u \dfrac{dI}{dz} + \alpha(z) I(z, u) = \alpha(z) B(z, u) \\ I(0, u) = 0 \quad \text{for } u > 0 \\ I(z_0, u) = B_s \quad \text{for } u < 0 \end{cases} \tag{6.7}$$

which is similar to that of the forward problem Eq. (4.13). Then, the solution $I(z, u)$ can be represented in the form, analogous to Eq. (4.18):

$$I(z, u) = -\frac{1}{u} \int_z^{z_0} B(z', u) \exp\left(\frac{1}{u} \int_z^{z'} \alpha(z'') dz''\right) \alpha(z') dz' + B_s \exp\left(\frac{1}{u} \int_z^{z_0} \alpha(z') dz'\right)$$

$$(6.8)$$

Then, the observable $R(\mu) = I(z, -\mu)$ takes the form analogous to Eq. (4.19):

$$R(\mu) = \frac{1}{\mu} \int_0^{z_0} B(z, -\mu) t(z, \mu) \alpha(z) dz + t(z_0, \mu) B_s \qquad (6.9)$$

where the atmospheric transmittance $t(z, \mu)$ has the form analogous to Eq. (4.20):

$$t(z, \mu) = \exp\left(-\frac{1}{\mu} \int_0^z \alpha(z') dz'\right) \qquad (6.10)$$

The similarities of the obtained expressions for the solution of the baseline forward problem Eq. (6.8), observable Eq. (6.9) and transmittance, Eq. (6.10) will be exploited below by re-use of results obtained in Chap. 5.

6.1.2 Linearization Approach

We linearize the baseline forward problem using its form of Eq. (6.7)

$$\begin{cases} \delta\left(u\dfrac{dI}{dz} + \alpha(z)I(z, u)\right) = \delta[\alpha(z)B(z, u)] \\ \delta I(0, u) = 0 \quad \text{for } u > 0 \\ \delta I(z_0, u) = \delta B_s \quad \text{for } u < 0 \end{cases} \qquad (6.11)$$

The arguments z and u will be suppressed for brevity. The equation of the linearized forward problem Eq. (6.11) can be rewritten in the form

$$u\frac{d\delta I}{dz} + \alpha\delta I = (B - I)\delta\alpha + \alpha\delta B \qquad (6.12)$$

Linearizing the expression for the atmospheric source function $B(z, u)$, Eq. (6.5), we have:

$$\delta B = \frac{1}{2} \int_{-1}^{1} \delta pI \ du' + \frac{1}{2} \int_{-1}^{1} p\delta I \ du' + (1 - \omega_0)\delta B - B\delta\omega_0 \qquad (6.13)$$

Substituting in Eq. (6.12) and re-grouping the terms, we obtain:

$$u\frac{d\delta I}{dz} + \alpha\left(\delta I - \frac{1}{2}\int_{-1}^{1} p\delta I \ du'\right) = (B - I)\delta\alpha$$

$$+ \alpha\left(\frac{1}{2}\int_{-1}^{1} \delta pI \ du' + (1 - \omega_0)\delta B - B\delta\omega_0\right)$$

$$(6.14)$$

Similarly, linearizing the expression for the surface source function B_s, Eq. (6.6), we have:

$$\delta B_s = 2\delta A \int_{0}^{1} Iu' \ du' + 2A \int_{0}^{1} \delta Iu' \ du' - \delta AB_s + (1 - A)\delta B_s \qquad (6.15)$$

Substituting Eq. (6.15) in the boundary condition of Eq. (6.11) and re-grouping the terms, we obtain the lower boundary condition in the form:

$$\delta I - 2A \int_{0}^{1} \delta Iu' \ du' = \left(2\int_{0}^{1} Iu' \ du' - B_s\right)\delta A + (1 - A)\delta B_s \qquad (6.16)$$

Thus, the resulting linearized forward problem obtains the form:

$$\begin{cases} L_e\delta I = (B - I)\delta\alpha + \alpha\left(\dfrac{1}{2}\displaystyle\int_{-1}^{1} \delta pI \ du' + (1 - \omega_0)\delta B - B\delta\omega_0\right) \\[4mm] L_t\delta I = 0 \quad \text{for } z = 0 \text{ and } u > 0 \\[4mm] L_b\delta I = \left(2\displaystyle\int_{0}^{1} Iu' \ du' - B_s\right)\delta A + (1 - A)\delta B_s \text{ for } z = z_0 \text{ and } u < 0 \end{cases} \qquad (6.17)$$

where the left-hand terms

$$L_e \delta I = u \frac{d\delta I}{dz} + \alpha \left(\delta I - \frac{1}{2} \int_{-1}^{1} p \delta I \, du' \right)$$

$$L_t \delta I = \delta I \qquad\qquad (6.18)$$

$$L_b \delta I = \delta I - 2A \int_{0}^{1} \delta I u' \, du'$$

contain the same linear operations as in the left-hand terms of the baseline forward problem, Eq. (6.1). Note that these operations do not depend on variations of specific parameters, all of which are contained in the right-hand terms of the equation and lower boundary condition. Also note that we re-use the baseline solution $I(z, u)$ as well as the atmospheric source function $B(z, u)$ of the baseline forward problem Eq. (6.5).

The baseline forward problem Eq. (6.1) contains four model input parameters – two discrete parameters: the surface albedo A and the surface Planck function B_s, and two continuous parameters: atmospheric extinction coefficient $\alpha(z)$, and the phase function of atmospheric scattering $p(z; u, u')$. Single scattering albedo $\omega_0(z)$ is not an independent parameter, but rather the zero-th expansion coefficient of $p(z; u, u')$ over Legendre polynomials in Eq. (6.2). By the same token, $p(z; u, u')$ should not be treated as a single continuous parameter, but rather a set of continuous parameters $\{\omega_l(z)\}$.

We first consider linearization with respect to the discrete parameters. Assuming that the only non-zero variation is that of the surface albedo, δA, we rewrite the linearized problem, Eq. (6.17) in the form:

$$\begin{cases} u \dfrac{d\delta_A I}{dz} + \alpha \left(\delta_A I - \dfrac{1}{2} \int_{-1}^{1} p \delta_A I \, du' \right) = 0 \\[2ex] \delta_A I = 0 \quad \text{for } z = 0 \text{ and } u > 0 \qquad\qquad (6.19) \\[2ex] \delta_A I - 2A \int_{0}^{1} \delta_A I u' \, du' = \left(2 \int_{0}^{1} I u' \, du' - B_s \right) \delta A \quad \text{for } z = z_0 \text{ and } u < 0 \end{cases}$$

Expressing the variation $\delta_A I$ through corresponding partial derivative $\partial I / \partial A \equiv I'_A$, and cancelling out the variation δA in both sides of the resulting equation and boundary conditions, we obtain the corresponding linearized forward problem:

$$\begin{cases} L_e I_A' = 0 \\ L_t I_A' = 0 \text{ for } z = 0 \text{ and } u > 0 \\ L_b I_A' = 2 \int_0^1 I u'\, du' - B_s \text{ for } z = z_0 \text{ and } u < 0 \end{cases} \tag{6.20}$$

Now, assuming that the only non-zero variation in Eq. (6.17) is that of the surface Planck function δB_s, denoting $I_{B_s}' \equiv \partial I / \partial B_s$, and acting in the same fashion as above for the surface albedo A, we obtain a corresponding linearized problem in the form:

$$\begin{cases} L_e I_{B_s}' = 0 \\ L_t I_{B_s}' = 0 \quad \text{for } z = 0 \text{ and } u > 0 \\ L_b I_{B_s}' = (1 - A)\delta B_s \quad \text{for } z = z_0 \text{ and } u < 0 \end{cases} \tag{6.21}$$

For continuous parameters we need to take variational derivatives of both sides of the equation and boundary conditions of the corresponding linearized problems. Assuming that the only non-zero variation is that of the atmospheric Planck function, $\delta B(z)$, we rewrite the linearized problem, Eq. (6.17) in the form:

$$\begin{cases} u \dfrac{d\delta_B I}{dz} + \alpha \left(\delta_B I - \dfrac{1}{2} \int_{-1}^1 p \delta_B I\, du' \right) = \alpha(1 - \omega_0)\delta B \\ \delta_B I = 0 \quad \text{for } z = 0 \text{ and } u > 0 \\ \delta_B I - 2A \int_0^1 \delta_B I u'\, du' = 0 \quad \text{for } z = z_0 \text{ and } u < 0 \end{cases} \tag{6.22}$$

Applying to the equation of the system, Eq. (6.22), the techniques for converting the variational equation Eq. (2.15) into the equation for variational derivatives Eq. (2.24), we obtain the linearized forward problem for the variational derivative (denoted by the "~" sign) $\delta I(z, u)/\delta B(z') \equiv I_B^{\sim}$:

$$\begin{cases} L_e I_B^{\sim} = \delta(z' - z) \cdot \alpha(z)(1 - \omega_0(z)) \\ L_t I_B^{\sim} = 0 \quad \text{for } u > 0 \\ L_b I_B^{\sim} = 0 \quad \text{for } u < 0 \end{cases} \tag{6.23}$$

Further on, assuming that the only non-zero variation is that of the atmospheric extinction coefficient, $\delta\alpha(z)$, we can rewrite the linearized problem, Eq. (6.17) in the form:

$$\begin{cases} L_e\delta_\alpha I = (\mathbf{B} - I)\delta\alpha \\ L_t\delta_\alpha I = 0 \quad \text{for } z = 0 \text{ and } u > 0 \\ L_b\delta_\alpha I = 0 \quad \text{for } z = z_0 \text{ and } u < 0 \end{cases} \tag{6.24}$$

Applying to the equation of the system, Eq. (6.24), the techniques for converting the variational equation Eq. (2.15) into the equation for variational derivatives Eq. (2.24) we obtain the linearized forward problem for the variational derivative $\delta I(z, u)/\delta\alpha(z') \equiv I_\alpha^\sim$:

$$\begin{cases} L_e I_\alpha^\sim = \delta(z' - z) \cdot (\mathbf{B}(z, u) - I(z, u)) \\ L_t I_\alpha^\sim = 0 \quad \text{for } u > 0 \\ L_b I_\alpha^\sim = 0 \quad \text{for } u < 0 \end{cases} \tag{6.25}$$

The last continuous parameter of the baseline forward problem, Eq. (6.1), is the phase function of atmospheric scattering $p(z, u, u')$, represented by its expansion over Legendre polynomials, Eq. (6.2), with coefficients $\omega_l(z)$. Assuming that the variation of the phase function δp (including $\delta\omega_0$) is the only variation in Eq. (6.17), we have:

$$\begin{cases} L_e\delta_p I = \alpha\left(\dfrac{1}{2}\displaystyle\int_{-1}^{1} \delta p I \, \mathrm{d}u' - \mathbf{B}\delta\omega_0\right) \\ L_t\delta_p I = 0 \quad \text{for } z = 0 \text{ and } u > 0 \\ L_b\delta_p I = 0 \quad \text{for } z = z_0 \text{ and } u < 0 \end{cases} \tag{6.26}$$

Substituting the varied expansion, Eq. (6.2),

$$\delta p(z; u, u') = \sum_{l=0}^{N} \delta\omega_l(z)P_l(u)P_l(u') \tag{6.27}$$

into the integral term in the equation of the system Eq. (6.26), we have:

$$\int_{-1}^{1} \delta p(z; u, u')I(z, u') \, \mathrm{d}u' = \sum_{l=0}^{N} \delta\omega_l(z)P_l(u) \cdot \int_{-1}^{1} I(z, u')P_l(u') \, \mathrm{d}u' \tag{6.28}$$

Assuming that the variation δp is due to an individual variation of the expansion coefficient $\delta\omega_l$, we obtain the linearized forward problem for the variational derivative $\delta I(z, u)/\delta\omega_l(z') \equiv I_{\omega_l}^\sim$:

$$\begin{cases} L_e I_{\omega_l}^{\sim} = \delta(z'-z) \cdot \alpha(z) \left(\frac{1}{2} P_l(u) \left(\int\limits_{-1}^{1} I(z,u') P_l(u') \, du' \right) - \delta_{0l} B(z) \right) \\ L_t I_{\omega_l}^{\sim} = 0 \quad \text{for } u > 0 \\ L_b I_{\omega_l}^{\sim} = 0 \quad \text{for } u < 0 \end{cases} \tag{6.29}$$

where δ_{0l} is the Kronecker delta-factor, which engages the second term in parentheses of the equation of the system Eq. (6.29) only for $l = 0$.

We have formulated the linearized forward problems for all of the model parameters of the baseline forward problem Eq. (6.1). Solutions of these problems are further used to compute corresponding derivatives of the observable, $R(\mu)$, given by Eq. (4.14), by direct substitution of obtained derivatives of the forward solution. It worth emphasizing that these linearized forward problems can be solved by the same methods as the baseline forward problem Eq. (6.1).

6.1.3 Adjoint Approach

Now we consider the application of the adjoint approach to the numerical model specified by the forward problem Eq. (6.1) and observable Eq. (6.9). To apply the Lagrange identity Eq. (3.20) and obtain the adjoint operator L^* of the corresponding adjoint problem, we build the corresponding linear operator and right-hand term of the operator equation Eq. (3.18). From Eq. (6.1) we have:

$$LI = u \frac{dI}{dz} + \alpha(z) \left(I - \frac{1}{2} \int\limits_{-1}^{1} pI \, du' \right) + \delta(z)\lambda(u)I + \delta(z-z_0)\lambda(-u) \left(I - 2A \int\limits_{0}^{1} Iu' du' \right) \tag{6.30}$$

Applying the Lagrange identity, we have:

$$\begin{aligned} (I^*, LI) &= \int\limits_{0}^{z_0} dz \int\limits_{-1}^{1} du I^* \left[u \frac{dI}{dz} + \alpha \left(I - \frac{1}{2} \int\limits_{-1}^{1} pI \, du' \right) + \delta(z)\lambda(u)I \right. \\ &\quad \left. + \delta(z-z_0)\lambda(-u) \left(I - 2A \int\limits_{0}^{1} Iu' du' \right) \right] \\ &= \int\limits_{-1}^{1} du \cdot u \int\limits_{0}^{z_0} dz I^* \frac{dI}{dz} + \int\limits_{0}^{z_0} dz \int\limits_{-1}^{1} du I^* \alpha \left(I - \frac{1}{2} \int\limits_{-1}^{1} pI \, du' \right) \\ &\quad + \int\limits_{-1}^{1} du \cdot u I^* I \big|_{z=0} - \int\limits_{-1}^{1} du \cdot u I^* I \big|_{z=z_0} + 2A \int\limits_{-1}^{0} du \cdot u I^* \big|_{z=z_0} \cdot \int\limits_{0}^{1} du' \cdot u' I \big|_{z=z_0} \end{aligned}$$

The inner integral in the first double integral term is integrated by parts:

$$\int_0^{z_0} I^* \frac{dI}{dz}\, dz = I^* I\big|_0^{z_0} - \int_0^{z_0} \frac{dI^*}{dz} I\, dz,$$

Then we exchange the inner integrals over u and u' in the second double integral term:

$$\int_0^{z_0} dz \int_{-1}^1 du I^* \alpha \left(I - \frac{1}{2} \int_{-1}^1 pI\, du' \right) = \int_0^{z_0} dz \int_{-1}^1 du \alpha \left(I^* - \frac{1}{2} \int_{-1}^1 pI^*\, du' \right) I,$$

and rewrite the term corresponding to the lower boundary condition as

$$2A \int_{-1}^0 du \cdot u I^*\big|_{z=z_0} \cdot \int_0^1 du' \cdot u' I\big|_{z=z_0} = \int_0^1 du \cdot u \left(2A \int_0^1 du' \cdot u' I^*\big|_{z=z_0} \right) I\big|_{z=z_0}$$

The resulting expression for (I^*, LI) can be rewritten in the form of (L^*I^*, I), where

$$L^*I^* = -u\frac{dI^*}{dz} + \alpha \left(I^* - \frac{1}{2} \int_{-1}^1 pI^*\, du' \right) + \delta(z)\lambda(-u)I^*$$

$$+ \delta(z - z_0)\lambda(u) \left(I^* + 2A \int_{-1}^0 I^* u'\, du' \right) \tag{6.31}$$

is the definition of the resulting adjoint operator.

We will use the expression for the observable in the same form Eq. (5.103) as in Chap. 5. This yields the right-hand term W in the form of Eq. (5.109) with components defined by Eqs. (5.111) and (5.113). Then, the resulting adjoint problem has the form:

$$\begin{cases} -u\dfrac{dI^*}{dz} + \alpha(z)\left(I^*(z, u) - \dfrac{1}{2} \int_{-1}^1 p(z; u, u')I^*(z, u')du' \right) = 0 \\[3mm] I^*(0, u) = \dfrac{1}{\mu}\delta(u + \mu), \quad \text{for } u > 0 \\[3mm] I^*(z_0, u) - 2A \int_0^1 I^*(z_0, u')u'du' = 0, \quad \text{for } u < 0 \end{cases} \tag{6.32}$$

Note that the adjoint problem Eq. (6.32) looks very similar to the forward problem Eq. (6.1) and can, therefore, be solved by the same numerical methods as Eq. (6.1).

Once the adjoint solution $I^*(z, u)$ is available numerically, it can be used to compute sensitivities of the observable $R(\mu)$ to discrete and continuous parameters of the forward problem Eq. (6.1). Here we use, correspondingly, the general expressions Eqs. (3.25) and (3.26) with the forward operator L specified by Eq. (6.30) and the right-hand term S assembled in a similar fashion from the right-hand terms of the equation and boundary conditions of Eq. (6.1):

$$S = \alpha(z)[1 - \omega_0(z)]B(z) + \delta(z)\lambda(u) \cdot 0 + \delta(z - z_0)\lambda(-u)(1 - A) \cdot B_s \quad (6.33)$$

To obtain the sensitivities to the discrete parameters of the baseline forward problem, Eq. (6.1), we use the general expression, Eq. (3.25). For the sensitivity to the surface Planck function B_s we have:

$$\frac{\partial L}{\partial B_s} I = 0; \quad \frac{\partial S}{\partial B_s} = \delta(z - z_0)(-u)\theta(-u)(1 - A)$$

Substituting in Eq. (3.25) we obtain:

$$\frac{\partial R(\mu)}{\partial B_s} = \int_{-1}^{0} I^*(z_0, u)(-u) \, du \cdot (1 - A) \quad (6.34)$$

Note that in Eq. (6.34), and in other expressions for sensitivities below, the value of $-u$ is non-negative in the whole interval of integration $u \in [-1, 0]$. Also, in the non-scattering limit, with $\omega_0(z) \to 0$, the adjoint solution $I^*(z, u)$ converges to Eq. (5.115), and correspondingly, Eq. (6.34) converges to Eq. (5.117).

For the sensitivity to the surface albedo A we have:

$$\frac{\partial S}{\partial A} - \frac{\partial L}{\partial A} I = \left(2 \int_{0}^{1} Iu' \, du' - B_s \right) \cdot \delta(z - z_0)(-u)\theta(-u)$$

Substituting in Eq. (3.25), we obtain:

$$\frac{\partial R(\mu)}{\partial A} = \int_{-1}^{0} I^*(z_0, u)(-u) \, du \cdot \left(2 \int_{0}^{1} I(z_0, u)u \, du - B_s \right) \quad (6.35)$$

Note that there is no direct convergence of Eq. (6.35) to Eq. (5.118) because of an extra term corresponding to the illumination of the scattering underlying surface by downwelling radiation, which was neglected in the lower boundary condition of the baseline forward problem Eq. (4.13).

To obtain the sensitivities to continuous parameters $B(z)$ and $\kappa(z)$ we use Eq. (3.26). For the atmospheric Planck function $B(z)$ from Eq. (6.30) follows that $\delta_B LI = 0$, and from Eq. (6.33) follows that $\delta_B S = \alpha(1 - \omega_0)\delta B = \kappa \delta B$. Therefore we have:

$$\frac{\delta L}{\delta B(z')} I = 0, \quad \frac{\delta S(z)}{\delta B(z')} = \delta(z' - z)\kappa(z)$$

Substituting in Eq. (3.26), we obtain:

$$\frac{\delta R(\mu)}{\delta B(z)} = \int\limits_0^{z_0} dz \int\limits_{-1}^1 du I^*(z', u) \cdot \frac{\delta S(z)}{\delta B(z')} = \kappa(z) \int\limits_{-1}^1 I^*(z', u)\, du \qquad (6.36)$$

For the extinction coefficient $\alpha(z)$, from Eqs. (6.30) and (6.33) it follows that

$$\delta_\alpha LI = \left(I - \frac{1}{2} \int\limits_{-1}^1 pI\, du'\right)\delta\alpha, \quad \delta_\alpha S = (1 - \omega_0)B\delta\alpha \qquad (6.37)$$

Therefore, we have:

$$\frac{\delta L}{\delta\alpha(z')} I = \delta(z' - z)\left(I - \frac{1}{2}\int\limits_{-1}^1 pI\, du'\right), \quad \frac{\delta S(z)}{\delta\alpha(z')} = \delta(z' - z)[1 - \omega_0(z)]B(z) \quad (6.38)$$

Combining these two terms and recalling the definition Eq. (6.5) of the atmospheric source function $B(z, u)$, we have:

$$\frac{\delta S(z)}{\delta\alpha(z')} - \frac{\delta L}{\delta\alpha(z')} I = \delta(z' - z)\left[\frac{1}{2}\int\limits_{-1}^1 pI\, du' + (1 - \omega_0)B - I\right]$$
$$= \delta(z' - z)[B(z, u) - I(z, u)] \qquad (6.39)$$

Substituting Eq. (6.39) in Eq. (3.26), we obtain:

$$\frac{\delta R(\mu)}{\delta\alpha(z)} = \int\limits_0^{z_0} dz \int\limits_{-1}^1 du I^*(z', u)\left(\frac{\delta S(z)}{\delta\kappa(z')} - \frac{\delta L}{\delta\kappa(z')} I(z', u)\right)$$
$$= \int\limits_0^{z_0} dz' \delta(z' - z) \int\limits_{-1}^1 du I^*(z', u)[B(z) - I(z, u)]$$
$$= \int\limits_{-1}^1 I^*(z, u)[B(z) - I(z, u)]\, du \qquad (6.40)$$

To obtain the sensitivities to expansion coefficients $\{\omega_l(z)\}$ of the phase function $p(z, u)$, we can use a different approach. Applying the general expression Eq. (3.24), we have:

$$\delta_p R(\mu) = \left(I^*, \delta_p S - \delta_p L I \right) \tag{6.41}$$

Varying the phase function in Eqs. (6.30) and (6.33), and combining the results, we have:

$$\delta_p S - \delta_p L I = \frac{1}{2} \int\limits_{-1}^{1} \delta p(z; u, u') I(z, u') \, \mathrm{d}u' - \delta \omega_0(z) B(z) \tag{6.42}$$

Using the expansion Eq. (6.2) of the phase function over Legendre polynomials, the first term in the right side of Eq. (6.42) can be rewritten as

$$\frac{1}{2} \int\limits_{-1}^{1} \delta p(z; u, u') I(z, u') \, \mathrm{d}u' = \frac{1}{2} \sum_{l=0}^{N} \delta \omega_l(z) P_l(u) \int\limits_{-1}^{1} I(z, u') P_l(u') \, \mathrm{d}u' \tag{6.43}$$

Observing that $P_0(u) \equiv 1$, and $\int_{-1}^{1} P_l(u) \, \mathrm{d}u = 0$ for all $l > 0$, the second term in the right side of Eq. (6.42) can be transformed to a similar form:

$$\delta \omega_0(z) B(z) = \frac{1}{2} \delta \omega_0(z) P_0(u) \int\limits_{-1}^{1} B(z) P_0(u') \, \mathrm{d}u'$$

$$= \sum_{l=0}^{N} \delta \omega_l(z) P_l(u) \int\limits_{-1}^{1} B(z) P_l(u') \, \mathrm{d}u' \tag{6.44}$$

Substituting the obtained expressions, Eqs. (6.43) and (6.44) in Eq. (6.42), and substituting the result in Eq. (6.41) we have:

$$\delta_p R(\mu) = \frac{1}{2} \int\limits_{-1}^{1} \mathrm{d}z \alpha(z) \sum_{l=0}^{N} \delta \omega_l(z) \left(\int\limits_{-1}^{1} I^*(z, u) P_l(u) \, \mathrm{d}u \right) \cdot \left\{ \int\limits_{-1}^{1} [I(z, u) - B(z)] P_l(u) \, \mathrm{d}u \right\} \tag{6.45}$$

Thus, $\delta_p R(\mu)$ can be represented in a form of a series

$$\delta_p R(\mu) = \sum_{l=0}^{N} \delta_{\omega_l} R(\mu) \tag{6.46}$$

with terms corresponding to separate variations of expansion coefficients $\delta\omega_l(z)$:

$$\delta_{\omega_l} R(\mu) = \frac{1}{2} \int\limits_{-1}^{1} dz \delta\omega_l(z) \alpha(z) \left(\int\limits_{-1}^{1} I^*(z,u) P_l(u) \, du \right) \cdot \left\{ \int\limits_{-1}^{1} [I(z,u') - B(z)] P_l(u') \, du' \right\}$$

(6.47)

Comparing with the general definition of the variational derivative, Eqs. (2.5), (2.6) we obtain the resulting expression for the sensitivities of the observable $R(\mu)$ to the expansion coefficients $\omega_l(z)$ of the phase function of atmospheric scattering in the form:

$$\frac{\delta R(\mu)}{\delta\omega_l(z)} = \alpha(z) \left(\int\limits_{-1}^{1} I^*(z,u) P_l(u) \, du \right) \cdot \left\{ \int\limits_{-1}^{1} [I(z,u') - B(z)] P_l(u') \, du' \right\}$$ (6.48)

Of course, there is no non-scattering analog to Eq. (6.48) because in this case all $\omega_l = 0$.

6.2 Zero-Dimensional Model of Atmospheric Dynamics

All examples of the application of sensitivity analysis considered above were done for models based on scalar differential equations for scalar functions of a single argument. In the remainder of this chapter we consider two examples of models employing the matrix differential equations for vector functions of a single argument. In the next chapter we will make a next step and consider the models employing matrix differential equations for vector functions of more than one argument.

6.2.1 Baseline Forward Problem and Observables

We consider a simplistic atmospheric model with the vector of forward solution consisting of just two components: surface temperature T and cloudiness n at the tropopause:

$$\mathbf{X}(t) = \begin{pmatrix} T(t) \\ n(t) \end{pmatrix}$$

(6.49)

The forward problem of this model has the form:

$$
\begin{cases}
c_p \dfrac{dT}{dt} + \left(1 - \dfrac{n}{2}\right)\sigma T^4 = (1 - nA)E_S(t) \\[2mm]
\tau \dfrac{dn}{dt} + n = \dfrac{T - T^{(0)}}{\Delta T} \\[2mm]
T(t_0) = T_0 \\[2mm]
n(t_0) = n_0
\end{cases}
\tag{6.50}
$$

The first equation of the system Eq. (6.50) describes the radiative balance between solar heating, radiative cooling, and the rate of resulting heating/cooling of the atmosphere. If the cloudiness $n = 0$, then the radiative balance is defined by solar heating E_S and radiative cooling σT^4 of the planetary surface. If the cloudiness $n = 1$, then the radiative balance is defined by solar heating $(1 - A)E_S$ and radiative cooling $\sigma T^4/2$ of the cloud tops at the tropopause having albedo A.

The second scalar equation of the system Eq. (6.50) describes the balance between the generation of cloudiness specified in the right-hand term by scaled excess temperature $(T - T^{(0)})/\Delta T$, and the decay of the cloudiness with a given time constant τ. Here, $T^{(0)}$ and ΔT are empirical model parameters.

The system, Eq. (6.50) can be re-written in matrix-vector form as:

$$
\begin{cases}
\mathbf{N}_e[\mathbf{X}] = \mathbf{S}_e \\
\mathbf{X}(t_0) = \mathbf{S}_c
\end{cases}
\tag{6.51}
$$

Here the non-linear 2×2 matrix differential operator of the equation in the system Eq. (6.51) has the form:

$$
\mathbf{N}_e[\mathbf{X}] = \frac{d}{dt}\begin{pmatrix} T \\ n \end{pmatrix} + \begin{pmatrix} \frac{1}{c_p}\left(1 - \frac{n}{2}\right)\sigma T^4 + \frac{nAE_S}{c_p} \\ -\frac{1}{\tau \Delta T}T + \frac{1}{\tau}n \end{pmatrix}
\tag{6.52}
$$

The right-hand terms of the equation and initial-value condition of the system Eq. (6.51) are correspondingly:

$$
\mathbf{S}_e = \begin{pmatrix} E_S/c_p \\ T^{(0)}/(\tau \Delta T) \end{pmatrix}, \quad \mathbf{S}_c = \begin{pmatrix} T_0 \\ n_0 \end{pmatrix} \equiv \mathbf{X}_0
\tag{6.53}
$$

The subscripts 'e' and 'c' denote, respectively, for "equation" and "condition".

In this model we assume a 2-vector of observables in the form:

$$
\mathbf{R} = \begin{pmatrix} T(t_1) \\ n(t_1) \end{pmatrix} \equiv \mathbf{X}(t_1)
\tag{6.54}
$$

which can be represented in the form of an inner product, or a convolution

$$\mathbf{R} = (\mathbf{W}, \mathbf{X}) = \int_{t_0}^{t_1} \mathbf{W}^T(t)\mathbf{X}(t) \, dt \qquad (6.55)$$

The corresponding observables weighting function has the form

$$\mathbf{W}(t) = \delta(t - t_1)\mathbf{I} \qquad (6.56)$$

where $\mathbf{I}$ is an 2×2 identity matrix and $\delta(t)$ is the Dirac delta function. We consider the sensitivity analysis of the above model with respect to three model parameters: the albedo of clouds A, and initial values of temperature T_0 and cloudiness n_0 represented in the form of a 2-vector $\mathbf{X}_0$.

6.2.2 Linearization Approach

Assuming arbitrary variations δA and $\delta \mathbf{X}_0$ of chosen model parameters, the system Eq. (6.51) can be linearized around the baseline solution and presented in the matrix-vector form as follows:

$$\begin{cases} \mathbf{L}_e \delta \mathbf{X} = \delta \mathbf{S}_e - \delta \mathbf{N}_e[\mathbf{X}] \\ \mathbf{L}_c \delta \mathbf{X} = \delta \mathbf{S}_c \text{ at } t = t_0 \end{cases} \qquad (6.57)$$

Here, $\mathbf{L}_e$ is a linear 2×2 matrix differential operator

$$\mathbf{L}_e \delta \mathbf{X} = \frac{d\delta \mathbf{X}}{dt} + \mathbf{C}(t)\delta \mathbf{X}(t) \qquad (6.58)$$

where

$$\mathbf{C}(t) = \begin{pmatrix} \frac{4}{c_p}\left(1 - \frac{n}{2}\right)\sigma T^3 & \frac{1}{c_p}\left(AE_S - \frac{1}{2}\sigma T^4\right) \\ -\frac{1}{\tau \Delta T} & \frac{1}{\tau} \end{pmatrix} \qquad (6.59)$$

The terms in the right-hand side of equation in the system Eq. (6.57) are 2-vectors and have the form:

$$\delta \mathbf{S}_e = \mathbf{0}, \quad \delta \mathbf{N}_e[\mathbf{X}] = \begin{pmatrix} (nE_S/c_p)\delta A \\ 0 \end{pmatrix} \qquad (6.60)$$

The operator $\mathbf{L}_c$ of the initial-value condition of the system Eq. (6.57) is a 2×2 identity matrix:

$$\mathbf{L}_c = \mathbf{I} \tag{6.61}$$

and the right-hand term $\delta \mathbf{S}_c$ of the initial-value condition is a 2-vector:

$$\delta \mathbf{S}_c = \delta \mathbf{X}_0 \tag{6.62}$$

There are no non-linear operations in the initial-value condition of the system Eq. (6.57).

Varying the albedo A and expressing the variations $\delta_A \mathbf{X}$, $\delta_A \mathbf{N}[\mathbf{X}]$, and $\delta_A \mathbf{S}_e$ through corresponding partial derivatives, we have:

$$\delta_A \mathbf{X} = \frac{\partial \mathbf{X}}{\partial A} \delta A \tag{6.63}$$

$$\delta_A \mathbf{N}_e[\mathbf{X}] = \frac{\partial \mathbf{N}_e[\mathbf{X}]}{\partial A} \delta A, \ \ \delta_A \mathbf{S}_e = \frac{\partial \mathbf{S}_e}{\partial A} \delta A = \mathbf{0} \tag{6.64}$$

The partial derivatives in Eqs. (6.63) and (6.64) are 2-vectors:

$$\frac{\partial \mathbf{X}}{\partial A} = \begin{pmatrix} \partial T / \partial A \\ \partial n / \partial A \end{pmatrix} \tag{6.65}$$

$$\frac{\partial \mathbf{N}_e[\mathbf{X}]}{\partial A} = \begin{pmatrix} n E_S(t) / c_p \\ 0 \end{pmatrix}, \ \ \frac{\partial \mathbf{S}_e}{\partial A} = \mathbf{0} \tag{6.66}$$

Substituting Eqs. (6.63) and (6.64) into the system Eq. (6.57), and cancelling out the variations δA we obtain a corresponding linearized forward problem in a matrix-vector form:

$$\begin{cases} \mathbf{L}_e \dfrac{\partial \mathbf{X}}{\partial A} = - \dfrac{\partial \mathbf{N}_e[\mathbf{X}]}{\partial A} \\ \dfrac{\partial \mathbf{X}}{\partial A}\Bigg|_{t_0} = \mathbf{0} \end{cases} \tag{6.67}$$

Note that the albedo A and the initial values T_0 and n_0 are independent model parameters, and therefore $\delta_A \mathbf{S}_c = \delta_A \mathbf{X}_0 \equiv \mathbf{0}$.

Varying the 2-vector of initial values $\mathbf{X}_0$ and expressing the variations $\delta_{\mathbf{X}_0} \mathbf{X}$, $\delta_{\mathbf{X}_0} \mathbf{N}_e[\mathbf{X}]$, and $\delta_{\mathbf{X}_0} \mathbf{S}_c$ through corresponding partial derivatives, we have:

$$\delta_{\mathbf{X}_0} \mathbf{X} = \frac{\partial \mathbf{X}}{\partial \mathbf{X}_0} \delta \mathbf{X}_0 \tag{6.68}$$

$$\delta_{\mathbf{X}_0}\mathbf{N}_e[\mathbf{X}] = \frac{\partial \mathbf{N}_e[\mathbf{X}]}{\partial \mathbf{X}_0}\delta\mathbf{X}_0, \quad \delta_{\mathbf{X}_0}\mathbf{S}_c = \frac{\partial \mathbf{S}_c}{\partial \mathbf{X}_0}\delta\mathbf{X}_0 \tag{6.69}$$

Here, the partial derivatives $\partial\mathbf{X}/\partial\mathbf{X}_0$, $\partial\mathbf{N}_e[\mathbf{X}]/\partial\mathbf{X}_0$, and $\partial\mathbf{S}_c/\partial\mathbf{X}_0$ are 2×2 matrices:

$$\frac{\partial \mathbf{X}}{\partial \mathbf{X}_0} = \begin{pmatrix} \partial T/\partial T_0 & \partial T/\partial n_0 \\ \partial n/\partial T_0 & \partial n/\partial n_0 \end{pmatrix} \tag{6.70}$$

$$\frac{\partial \mathbf{N}_e[\mathbf{X}]}{\partial \mathbf{X}_0} = \begin{pmatrix} 0 & 0 \\ 0 & 0 \end{pmatrix} \equiv \mathbf{O}, \quad \frac{\partial \mathbf{S}_c}{\partial \mathbf{X}_0} = \begin{pmatrix} 1 & 0 \\ 0 & 1 \end{pmatrix} \equiv \mathbf{I} \tag{6.71}$$

The components of the 2-vector $\mathbf{N}_e[\mathbf{X}]$ do not depend on components of the vector of initial-value conditions $\mathbf{X}_0$. Therefore the corresponding matrix of partial derivatives reduces to a 2×2 null matrix $\mathbf{O}$. On the other hand, the 2-vector $\mathbf{S}_c \equiv \mathbf{X}_0$, and the corresponding matrix of partial derivatives reduces to a 2×2 identity matrix $\mathbf{I}$.

Substituting Eqs. (6.68) and (6.69) into the system Eq. (6.57) and cancelling out the variations $\delta\mathbf{X}_0$ yields a corresponding linearized forward problem in a matrix-vector form:

$$\begin{cases} \mathbf{L}_e \dfrac{\partial \mathbf{X}}{\partial \mathbf{X}_0} = \mathbf{O} \\ \dfrac{\partial \mathbf{X}}{\partial \mathbf{X}_0}\bigg|_{t_0} = \mathbf{I} \end{cases} \tag{6.72}$$

Expressions for corresponding sensitivities are obtained from Eqs. (6.55) and (6.56):

$$\frac{\partial \mathbf{R}}{\partial A} = \left(\mathbf{W}, \frac{\partial \mathbf{X}}{\partial A}\right) = \frac{\partial \mathbf{X}}{\partial A}\bigg|_{t=t_1} \tag{6.73}$$

and

$$\frac{\partial \mathbf{R}}{\partial \mathbf{X}_0} = \left(\mathbf{W}, \frac{\partial \mathbf{X}}{\partial \mathbf{X}_0}\right) = \frac{\partial \mathbf{X}}{\partial \mathbf{X}_0}\bigg|_{t=t_1} \tag{6.74}$$

These sensitivities have the form of a 2-vector and of a 2×2 matrix respectively.

6.2.3 Adjoint Approach

Now we consider the application of the adjoint approach to the numerical model specified by the forward problem Eq. (6.50) and observables weighting function

Eq. (6.56). To apply the Lagrange identity Eq. (3.20) and obtain the adjoint operator $\mathbf{L}^*$ of the corresponding adjoint problem, we need to formulate the corresponding linear operator of the operator equation Eq. (3.18). From the system Eq. (6.57) with the linear operator of the equation defined by Eq. (6.58), we have:

$$\mathbf{L}\delta\mathbf{X} = \frac{d\delta\mathbf{X}}{dt} + \mathbf{C}(t)\delta\mathbf{X}(t) + \delta(t - t_0)\delta\mathbf{X}(t) \tag{6.75}$$

The δ-function factor corresponds to the initial-value condition enforced at the instant $t = t_0$.

Applying the Lagrange identity to two arbitrary functions $\mathbf{X}(t)$ and $\mathbf{X}^*(t)$, we have:

$$(\mathbf{X}^*, \mathbf{L}\mathbf{X}) = \int_{t_0}^{t_1} \mathbf{X}^{*T} \left(\frac{d\mathbf{X}}{dt} + \mathbf{C}(t)\mathbf{X}(t) + \delta(t - t_0)\mathbf{X}(t) \right) \, dt \tag{6.76}$$

After integrating by parts and regrouping the terms, the resulting expression for the inner product $(\mathbf{X}^*, \mathbf{L}\mathbf{X})$ can be rewritten in the form:

$$(\mathbf{X}^*, \mathbf{L}\mathbf{X}) = \int_{t_0}^{t_1} \left(-\frac{d\mathbf{X}^*}{dt} + \mathbf{C}^T(t)\mathbf{X}^*(t) + \delta(t - t_1)\mathbf{X}^*(t) \right) \mathbf{X}(t) \, dt = (\mathbf{L}^*\mathbf{X}^*, \mathbf{X})$$

$$\tag{6.77}$$

where

$$\mathbf{L}^*\mathbf{X}^* = -\frac{d\mathbf{X}^*}{dt} + \mathbf{C}^T(t)\mathbf{X}^*(t) + \delta(t - t_1)\mathbf{X}^*(t) \tag{6.78}$$

is the resulting adjoint operator. To formulate the corresponding adjoint problem $\mathbf{L}^*\mathbf{X}^* = \mathbf{W}$, we rewrite the definition of the observables weighing function, Eq. (6.56), in the form:

$$\mathbf{W}(t) = \mathbf{W}_e(t) + \delta(t - t_1)\mathbf{W}_c \tag{6.79}$$

where $\mathbf{W}_e(t) \equiv \mathbf{O}$ and $\mathbf{W}_c = \mathbf{I}$. The resulting adjoint problem has the form of a final-value problem:

$$\begin{cases} -\dfrac{d\mathbf{X}^*}{dt} + \mathbf{C}^T(t)\mathbf{X}^*(t) = \mathbf{O} \\ \mathbf{X}^*(t_1) = \mathbf{I} \end{cases} \tag{6.80}$$

To obtain the sensitivities to the selected model parameters we use the general expression, Eq. (3.25). For the sensitivity to cloud albedo A, using Eq. (6.66) we have:

$$\frac{\partial \mathbf{R}}{\partial A} = \left(\mathbf{X}^*, -\frac{\partial \mathbf{N}_e[\mathbf{X}]}{\partial A} \right) = -\int_{t_0}^{t_1} \mathbf{X}^{*T}(t) \frac{\partial \mathbf{N}_e[\mathbf{X}]}{\partial A} \, dt \qquad (6.81)$$

In the expanded form we have:

$$\mathbf{X}^{*T}(t) \frac{\partial \mathbf{N}_e[\mathbf{X}]}{\partial A} = \begin{pmatrix} X_{11}^*(t) & X_{21}^*(t) \\ X_{12}^*(t) & X_{22}^*(t) \end{pmatrix} \begin{pmatrix} nE_S(t)/c_p \\ 0 \end{pmatrix} = \begin{pmatrix} X_{11}^*(t) \\ X_{12}^*(t) \end{pmatrix} \cdot \frac{nE_S(t)}{c_p} \qquad (6.82)$$

Thus

$$\begin{pmatrix} \partial T(t_1)/\partial A \\ \partial n(t_1)/\partial A \end{pmatrix} = -\frac{n}{c_p} \int_{t_0}^{t_1} \begin{pmatrix} X_{11}^*(t) \\ X_{12}^*(t) \end{pmatrix} nE_S(t) \, dt \qquad (6.83)$$

For the sensitivities to the components T_0 and n_0 of the initial-value vector $\mathbf{X}_0$, using Eqs. (6.71), we have:

$$\frac{\partial \mathbf{R}}{\partial \mathbf{X}_0} = \left(\mathbf{X}^*, \frac{\partial \mathbf{S}}{\partial \mathbf{X}_0} \right) = \int_{t_0}^{t_1} \mathbf{X}^{*T}(t) \delta(t - t_0) \frac{\partial \mathbf{S}_c}{\partial \mathbf{X}_0} \, dt = \mathbf{X}^{*T}(t_0)\mathbf{I} = \mathbf{X}^{*T}(t_0) \quad (6.84)$$

or, in the expanded form:

$$\begin{pmatrix} \partial T(t_1)/\partial T(t_0) & \partial T(t_1)/\partial n(t_0) \\ \partial n(t_1)/\partial T(t_0) & \partial n(t_1)/\partial n(t_0) \end{pmatrix} = \begin{pmatrix} X_{11}^*(t_0) & X_{21}^*(t_0) \\ X_{12}^*(t_0) & X_{22}^*(t_0) \end{pmatrix} \qquad (6.85)$$

6.3 Model of Orbital Tracking Data of the Planetary Orbiter Spacecraft

6.3.1 Baseline Forward Problem and Observables

In the model considered in this Section, the action occurs in the real 3D space. We consider the spacecraft orbiting a planet. The motion of the spacecraft is assumed to be defined solely by the gradient of potential of the planetary gravity field (no atmospheric drag and other interfering forces). The initial formulation of the baseline forward problem here has the form of an initial-value problem with the 2nd order ordinary differential equation:

$$\begin{cases} \dfrac{d^2\mathbf{r}}{dt^2} = \nabla U(\mathbf{r}) \\[2mm] \mathbf{r}\big|_{t_0} = \mathbf{r}_0, \ \dfrac{d\mathbf{r}}{dt}\bigg|_{t_0} = \mathbf{v}_0 \end{cases} \tag{6.86}$$

Here $U(\mathbf{r})$ is the gravity potential, gradient of which, ∇U, represents the local gravity acceleration, $\mathbf{r}$ is the radius vector of the spacecraft in planetocentric coordinates, and $\mathbf{r}_0$ and $\mathbf{v}_0$ are initial values of 3D position and velocity, also in planetocentric coordinates. In planetocentric coordinates, the gravity potential U at a given location $\mathbf{r}$ is represented by an expansion over the gravity harmonics J_n, C_{nm}, and S_{nm}:

$$\begin{aligned} U(\mathbf{r}) = U(r, \varphi, \lambda) = \frac{\mu}{r}\Bigg\{ &-1 + \sum_{n=2}^{N} \left(\frac{a_p}{r}\right)^n \Big[J_n P_n(\sin \varphi) \\ &+ \sum_{m=1}^{n} P_n^m(\sin \varphi)(C_{nm}\cos m\lambda + S_{nm}\sin m\lambda)\Big]\Bigg\} \end{aligned} \tag{6.87}$$

where r, φ, and λ are radial distance, latitude and longitude of the location, a_p and μ are the mean equatorial radius and gravitational constant of the planet, $P_n(\sin \varphi)$ are Legendre polynomials, and $P_n^m(\sin \varphi)$ are associated Legendre functions.

The 3-vector of the gradient of gravity potential ∇U has the form:

$$\begin{aligned} \nabla U = \begin{pmatrix} \frac{\partial U}{\partial r} \\[1mm] \frac{1}{r\cos\lambda}\cdot\frac{\partial U}{\partial \varphi} \\[1mm] \frac{1}{r}\cdot\frac{\partial U}{\partial \lambda} \end{pmatrix} = \frac{\mu}{r^2}\Bigg\{ &-\begin{pmatrix} 1 \\ 0 \\ 0 \end{pmatrix} + \sum_{n=2}^{N}\left(\frac{a_p}{r}\right)^n\Bigg[J_n\begin{pmatrix} (n+1)P_n \\ 0 \\ -\cos\varphi(P_n)' \end{pmatrix} \\ &+ \sum_{m=1}^{n}\begin{pmatrix} -(n+1)P_n^m(C_{nm}\cos m\lambda + S_{nm}\sin m\lambda) \\ m\sec\varphi P_n^m(-C_{nm}\cos m\lambda + S_{nm}\sin m\lambda) \\ \cos\varphi\left(P_n^m\right)'(C_{nm}\cos m\lambda + S_{nm}\sin m\lambda) \end{pmatrix}\Bigg]\Bigg\} \end{aligned} \tag{6.88}$$

This is the first problem in this book, where we encounter a differential equation of higher than first order. Obviously, this equation can be linearized in a straightforward way without any further transformations. But bearing in mind that later on we will formulate a corresponding adjoint problem, and as before, the elements of the linearized forward problem will be used for adjoint sensitivity analysis, we wish to transform the system Eq. (6.86) to that with a matrix differential equation of 1st order, which is more amenable to application of the Lagrange identity. In the next chapter, we will routinely apply this transformation to higher-order differential equations with partial derivatives.

Introducing the 3-vector of velocity $\mathbf{v}(t)$, we rewrite Eq. (6.86) in the form:

$$
\begin{cases}
\dfrac{d\mathbf{r}}{dt} - \mathbf{v}(t) = \mathbf{0} \\[2mm]
\dfrac{d\mathbf{v}}{dt} = \nabla U(\mathbf{r}) \\[2mm]
\mathbf{r}\big|_{t_0} = \mathbf{r}_0, \mathbf{v}\big|_{t_0} = \mathbf{v}_0
\end{cases}
\tag{6.89}
$$

Introducing a composite 6-vector

$$
\mathbf{X}(t) = \begin{pmatrix} \mathbf{r}(t) \\ \mathbf{v}(t) \end{pmatrix}
\tag{6.90}
$$

which combines the position and velocity of the spacecraft, we can rewrite Eq. (6.89) in the form of an initial-value problem with a matrix differential equation of 1st order:

$$
\begin{cases}
\dfrac{d\mathbf{X}}{dt} + \mathbf{A}\mathbf{X}(t) = \mathbf{B}(\mathbf{X}) \\[2mm]
\mathbf{X}(t_0) = \mathbf{X}_0
\end{cases}
\tag{6.91}
$$

Here

$$
\mathbf{A} = \begin{pmatrix} \mathbf{O} & -\mathbf{I} \\ \mathbf{O} & \mathbf{O} \end{pmatrix}
\tag{6.92}
$$

is a constant block matrix composed of four 3×3 matrices, where $\mathbf{O}$ is a null matrix and $\mathbf{I}$ is an identity matrix. The right-hand terms in the system Eq. (6.91) are composite 6-vectors:

$$
\mathbf{B}(\mathbf{X}) = \begin{pmatrix} \mathbf{0} \\ \nabla U(\mathbf{r}) \end{pmatrix}, \quad \mathbf{X}_0 = \begin{pmatrix} \mathbf{r}_0 \\ \mathbf{v}_0 \end{pmatrix}
\tag{6.93}
$$

In this model we assume that we deal with two instant observables: the geocentric distance to the spacecraft D, and its geocentric radial velocity V. We also assume that the geocentric position $\mathbf{r}_p$ and velocity $\mathbf{v}_p$ of the center of the planet are known. Then D and V are magnitudes (Euclidean norms) of sums of corresponding vectors:

$$
D = |\mathbf{r}_p + \mathbf{r}|, \quad V = |\mathbf{v}_p + \mathbf{v}|
\tag{6.94}
$$

Then, the 2-vector of observables in this model is defined as

$$\mathbf{R} = \begin{pmatrix} D \\ V \end{pmatrix} \tag{6.95}$$

Obviously, $\mathbf{R}$ is a non-linear vector function of vectors $\mathbf{r}$ and $\mathbf{v}$, and thus it is a non-linear function of the 6-vector $\mathbf{X}$:

$$\mathbf{R} = \mathbf{M}(\mathbf{X}) \tag{6.96}$$

In the applications below, we will need to linearize Eq. (6.96) with respect to $\mathbf{X}$. From Eq. (6.94) we have:

$$D^2 = (\mathbf{r}_p + \mathbf{r})^T (\mathbf{r}_p + \mathbf{r}), \quad V^2 = (\mathbf{v}_p + \mathbf{v})^T (\mathbf{v}_p + \mathbf{v}) \tag{6.97}$$

Assuming that variations $\delta(D^2)$ and $\delta(V^2)$ occur due to variations $\delta\mathbf{r}$ and $\delta\mathbf{v}$, we have:

$$\delta(D^2) = \delta\mathbf{r}^T (\mathbf{r}_p + \mathbf{r}) + (\mathbf{r}_p + \mathbf{r})^T \delta\mathbf{r} = 2(\mathbf{r}_p + \mathbf{r})^T \delta\mathbf{r} \tag{6.98}$$

$$\delta(V^2) = \delta\mathbf{v}^T (\mathbf{v}_p + \mathbf{v}) + (\mathbf{v}_p + \mathbf{v})^T \delta\mathbf{v} = 2(\mathbf{v}_p + \mathbf{v})^T \delta\mathbf{v} \tag{6.99}$$

On the other hand,

$$\delta(D^2) = 2D\delta D, \quad \delta(V^2) = 2V\delta V \tag{6.100}$$

Comparing Eqs. (6.98) and (6.99) with Eq. (6.100), we obtain:

$$\delta\mathbf{R} = \mathbf{W}_R^T \delta\mathbf{X}(t_R) \tag{6.101}$$

where

$$\mathbf{W}_R = \begin{pmatrix} \frac{1}{D}(\mathbf{r}_p + \mathbf{r}(t_R)) & \mathbf{0} \\ \mathbf{0} & \frac{1}{V}(\mathbf{v}_p + \mathbf{v}(t_R)) \end{pmatrix}\Bigg|_{t_R} \tag{6.102}$$

and t_R is an instant of the measurement. The matrix $\mathbf{W}_R$ is a 6×2 matrix.

We consider the sensitivity analysis of the model represented by Eqs. (6.91) and (6.102) with respect to the expansion coefficients of the gravity potential U, Eq. (6.87). The model parameters here are expansion coefficients $J_n(n = 2, \ldots N)$ and $C_{nm}, S_{nm}(n = 2, \ldots N, m = 1, \ldots n)$. For convenience of the derivations below, these expansion coefficients, hereafter referred to as gravity harmonics, are combined into an $N(N + 1)$ -vector $\mathbf{H}$.

6.3.2 Linearization Approach

Assuming an arbitrary variation of the gravity potential $U(\mathbf{r})$, we linearize Eq. (6.91) in the form [cf. Eqs. (6.92) and (6.93)]:

$$\begin{cases} \dfrac{d\delta\mathbf{X}}{dt} + \begin{pmatrix} \mathbf{O} & -\mathbf{I} \\ \mathbf{O} & \mathbf{O} \end{pmatrix} \delta\mathbf{X}(t) = \begin{pmatrix} \mathbf{0} \\ \delta[\nabla U(\mathbf{r})] \end{pmatrix} \\ \delta\mathbf{X}(t_0) = \mathbf{0} \end{cases} \tag{6.103}$$

Performing the transformation $\delta[\nabla U(\mathbf{r})] = \nabla[\delta U(\mathbf{r})]$, we further analyze the variation of the gravity potential $\delta U(\mathbf{r})$. It consists of two components: due to (a) variations of harmonics of the gravity field, and (b) variation of the trajectory of the spacecraft. We have respectively:

$$\delta[U(\mathbf{r})] = \delta_{\mathbf{H}} U + \delta_{\mathbf{r}} U(\mathbf{r}) \tag{6.104}$$

The second term in Eq. (6.104) has the form:

$$\delta_{\mathbf{r}} U(\mathbf{r}) = \frac{\partial U}{\partial x}\delta x + \frac{\partial U}{\partial y}\delta y + \frac{\partial U}{\partial z}\delta z = \nabla^T U \delta\mathbf{r} \tag{6.105}$$

and, we have:

$$\delta[\nabla U(\mathbf{r})] = \delta_{\mathbf{H}}\nabla U + \nabla\nabla^T U \delta\mathbf{r} \tag{6.106}$$

Substituting Eq. (6.106) in Eq. (6.103) and rearranging the terms, we obtain the linearized forward problem in a form:

$$\begin{cases} \dfrac{d\delta\mathbf{X}}{dt} + \mathbf{C}(t)\delta\mathbf{X}(t) = \delta\mathbf{B}(t) \\ \delta\mathbf{X}(t_0) = \mathbf{0} \end{cases} \tag{6.107}$$

where the 6×6 matrix $\mathbf{C}(t)$ and 6-vector $\delta\mathbf{B}(t)$ have the form:

$$\mathbf{C}(t) = \begin{pmatrix} \mathbf{O} & -\mathbf{I} \\ \nabla\nabla^T U(\mathbf{r}) & \mathbf{O} \end{pmatrix} \tag{6.108}$$

$$\delta\mathbf{B}(t) = \begin{pmatrix} \mathbf{0} \\ \delta_{\mathbf{H}}\nabla U(\mathbf{r}) \end{pmatrix} \tag{6.109}$$

It should be noted that the matrix $\mathbf{C}(t)$ and vector $\delta\mathbf{B}(t)$ are computed along the known trajectory $\mathbf{r}(t)$, which is the component of the baseline solution $\mathbf{X}(t)$, and therefore they are functions of time t only. Also, note that the variations of gravity harmonics are confined in the vector $\delta\mathbf{B}(t)$. From Eq. (6.88) we have:

$$\delta_{\mathbf{H}}\nabla U = \frac{\mu}{r^2}\sum_{n=2}^{N}\left(\frac{a_p}{r}\right)^n\left[\delta J_n\begin{pmatrix}(n+1)P_n\\0\\-\cos\varphi(P_n)'\end{pmatrix}+\sum_{m=1}^{n}\begin{pmatrix}-(n+1)P_n^m(\delta C_{nm}\cos m\lambda+\delta S_{nm}\sin m\lambda)\\m\sec\varphi P_n^m(-\delta C_{nm}\cos m\lambda+\delta S_{nm}\sin m\lambda)\\\cos\varphi(P_n^m)'(\delta C_{nm}\cos m\lambda+\delta S_{nm}\sin m\lambda)\end{pmatrix}\right]$$

$$(6.110)$$

Obtaining the expression for the 3×3 matrix $\nabla\nabla^T U$ requires performing some algebra. This matrix can be represented in the form of a row of three 3-vectors

$$\nabla\nabla^T U = \nabla\left(\frac{\partial U}{\partial x},\ \frac{\partial U}{\partial y},\ \frac{\partial U}{\partial z}\right) = \left(\frac{\partial\nabla U}{\partial r},\ \frac{1}{r\cos\varphi}\cdot\frac{\partial\nabla U}{\partial\lambda},\ \frac{1}{r}\cdot\frac{\partial\nabla U}{\partial\varphi}\right) \qquad (6.111)$$

where:

$$\frac{\partial\nabla U}{\partial r} = -\frac{3\mu}{r^3}\begin{pmatrix}1\\0\\0\end{pmatrix}+\frac{\mu}{r^3}\sum_{n=2}^{N}\left(\frac{a_p}{r}\right)^n J_n(n+2)\begin{pmatrix}(n+1)P_n\\0\\-\cos\varphi(P_n)'\end{pmatrix}$$
$$+\frac{\mu}{r^3}\sum_{n=2}^{N}\left(\frac{a_p}{r}\right)^n J_n(n+2)\sum_{m=1}^{n}\begin{pmatrix}-(n+1)P_n^m(C_{nm}\cos m\lambda+S_{nm}\sin m\lambda)\\m\sec\varphi P_n^m(-C_{nm}\cos m\lambda+S_{nm}\sin m\lambda)\\\cos\varphi(P_n^m)'(C_{nm}\cos m\lambda+S_{nm}\sin m\lambda)\end{pmatrix}$$

$$(6.112)$$

$$\frac{1}{r\cos\varphi}\cdot\frac{\partial\nabla U}{\partial\lambda} = \frac{\mu}{r^3}\sum_{n=2}^{N}\left(\frac{a_p}{r}\right)^n\sum_{m=1}^{n}m\begin{pmatrix}(n+1)P_n^m(C_{nm}\cos m\lambda+S_{nm}\sin m\lambda)\\m\sec^2\varphi P_n^m(-C_{nm}\cos m\lambda+S_{nm}\sin m\lambda)\\\cos\varphi(P_n^m)'(-C_{nm}\cos m\lambda+S_{nm}\sin m\lambda)\end{pmatrix}$$

$$(6.113)$$

$$\frac{1}{r}\cdot\frac{\partial\nabla U}{\partial\varphi} = -\frac{\mu}{r^3}\sum_{n=2}^{N}\left(\frac{a_p}{r}\right)^n J_n\begin{pmatrix}(n+1)\cos\varphi(P_n)'\\0\\\sin\varphi(P_n)'-\cos\varphi(P_n)''\end{pmatrix}$$
$$-\frac{\mu}{r^3}\sum_{n=2}^{N}\left(\frac{a_p}{r}\right)^n\sum_{m=1}^{n}\begin{pmatrix}-(n+1)\cos\varphi(P_n^m)''(C_{nm}\cos m\lambda+S_{nm}\sin m\lambda)\\m\left(\sin\varphi\sec^2\varphi P_n^m+(P_n^m)'\right)(-C_{nm}\cos m\lambda+S_{nm}\sin m\lambda)\\\left(\cos^2\varphi(P_n^m)''-\sin\varphi(P_n^m)'\right)(C_{nm}\cos m\lambda+S_{nm}\sin m\lambda)\end{pmatrix}$$

$$(6.114)$$

Now, expressing the variations $\delta_{\mathbf{H}}\mathbf{X}$ and $\delta_{\mathbf{H}}\nabla U$ through corresponding partial derivatives, we have:

$$\delta_{\mathbf{H}}\mathbf{X} = \frac{\partial\mathbf{X}}{\partial\mathbf{H}}\delta\mathbf{H} \qquad (6.115)$$

$$\delta_{\mathbf{H}} \nabla U = \frac{\partial \nabla U}{\partial \mathbf{H}} \delta \mathbf{H} \tag{6.116}$$

Here, the partial derivative $\partial \mathbf{X} / \partial \mathbf{H}$ is a matrix with 6 rows and $N(N+1)$ columns. The partial derivative $\partial \nabla U / \partial \mathbf{H}$ is a matrix with 3 rows and $N(N+1)$ columns. Therefore, the partial derivative $\partial \mathbf{B} / \partial \mathbf{H}$ [cf. Eq. (6.109)]:

$$\frac{\partial \mathbf{B}}{\partial \mathbf{H}} = \begin{pmatrix} \mathbf{O} \\ \partial \nabla U / \partial \mathbf{H} \end{pmatrix} \tag{6.117}$$

is a matrix with 6 rows and $N(N+1)$ columns. The elements of the matrix $\partial \nabla U / \partial \mathbf{H}$ are obtained from Eq. (6.88). We have:

$$\frac{\partial \nabla U}{\partial J_n} = \frac{\mu}{r^2} \left(\frac{a_p}{r}\right)^n \begin{pmatrix} (n+1)P_n \\ 0 \\ -\cos\varphi (P_n)' \end{pmatrix}, \quad n = 2, \dots N \tag{6.118}$$

$$\frac{\partial \nabla U}{\partial C_{nm}} = \frac{\mu}{r^2} \left(\frac{a_p}{r}\right)^n \begin{pmatrix} -(n+1)P_n^m \\ -m\sec\varphi P_n^m \\ \cos\varphi (P_n^m)' \end{pmatrix} \cos m\lambda, \quad \frac{\partial \nabla U}{\partial S_{nm}} = \frac{\mu}{r^2} \left(\frac{a_p}{r}\right)^n \begin{pmatrix} -(n+1)P_n^m \\ -m\sec\varphi P_n^m \\ \cos\varphi (P_n^m)' \end{pmatrix} \sin m\lambda,$$

$$n = 2, \dots N, \ m = 1, \dots n$$

$$\tag{6.119}$$

With all necessary inputs now available, we obtain the corresponding linearized forward problem in a matrix-vector form:

$$\begin{cases} \mathbf{L}_e \dfrac{\partial \mathbf{X}}{\partial \mathbf{H}} = \dfrac{\partial \mathbf{B}}{\partial \mathbf{H}} \\ \dfrac{\partial \mathbf{X}}{\partial \mathbf{H}} \bigg|_{t_0} = \mathbf{O} \end{cases} \tag{6.120}$$

where the operator of the equation has the form:

$$\mathbf{L}_e = \frac{\mathrm{d}}{\mathrm{d}t} + \mathbf{C}(t) \tag{6.121}$$

The expression for sensitivities with respect to gravity harmonics is obtained from Eq. (6.101):

$$\frac{\partial \mathbf{R}}{\partial \mathbf{H}} = \mathbf{W}_R^T \frac{\partial \mathbf{X}}{\partial \mathbf{H}} \bigg|_{t_R} \tag{6.122}$$

Recalling the definition of the matrix $\mathbf{W}_R$, Eq. (6.102), we see that the matrix of the sensitivities $\partial \mathbf{R} / \partial \mathbf{H}$ has the form of a matrix with 2 rows and $N(N+1)$ columns.

In the next Subsection, we will obtain the sensitivities to gravity harmonics using an alternative adjoint approach of sensitivity analysis.

6.3.3 Adjoint Approach

To apply the Lagrange identity Eq. (3.20) and obtain the adjoint operator $\mathbf{L}^*$ of the corresponding adjoint problem, we formulate the corresponding linear operator of the operator equation Eq. (3.18). From the system Eq. (6.107) we have:

$$\mathbf{L}\delta\mathbf{X} = \frac{\mathrm{d}\delta\mathbf{X}}{\mathrm{d}t} + \mathbf{C}(t)\delta\mathbf{X}(t) + \delta(t - t_0)\delta\mathbf{X}(t) \qquad (6.123)$$

Thus, in the matrix-vector form, Eq. (6.123) is identical to Eq. (6.35) of the previous Section, and we immediately obtain the adjoint operator in the form [cf. Eq. (6.78)]:

$$\mathbf{L}^*\mathbf{X}^* = -\frac{\mathrm{d}\mathbf{X}^*}{\mathrm{d}t} + \mathbf{C}^T(t)\mathbf{X}^*(t) + \delta(t - t_1)\mathbf{X}^*(t) \qquad (6.124)$$

We first formulate the corresponding adjoint problem $\mathbf{L}^*\mathbf{X}^* = \mathbf{W}$ for the simplest case of the observable $\mathbf{R}$ obtained at the end of the integration period $t_R = t_1$. We write the definition of the observables weighing function in the form [cf. Eq. (6.79)]:

$$\mathbf{W}(t) = \mathbf{W}_e(t) + \delta(t - t_1)\mathbf{W}_R \qquad (6.125)$$

Here $\mathbf{W}_e(t) \equiv \mathbf{O}$ is a null 6×2 matrix, and $\mathbf{W}_R$ is also a 6×2 matrix defined by Eq. (6.102), where $t_R = t_1$. The resulting adjoint problem has the form of a final-value problem:

$$\begin{cases} -\dfrac{\mathrm{d}\mathbf{X}^*}{\mathrm{d}t} + \mathbf{C}^T(t)\mathbf{X}^*(t) = \mathbf{O} \\ \mathbf{X}^*(t_1) = \mathbf{W}_R \end{cases} \qquad (6.126)$$

The solution $\mathbf{X}^*(t)$ is also a 6×2 matrix.

To obtain the matrix of sensitivities $\partial\mathbf{R}/\partial\mathbf{H}$ to the vector of gravity harmonics $\mathbf{H}$, we need to represent the right-hand term $\delta\mathbf{S}$ of the operator equation $\mathbf{L}\delta\mathbf{X} = \delta\mathbf{S}$ [cf. Eq. (3.18)] in the same form as the operator $\mathbf{L}\delta\mathbf{X}$. From the system Eq. (6.107) we have:

$$\delta\mathbf{S} = \delta\mathbf{B} + \delta(t - t_0) \cdot \mathbf{0} \qquad (6.127)$$

Using the general expression, Eq. (3.25), we obtain an expression for the sensitivities, which can be evaluated numerically:

$$\frac{\partial \mathbf{R}}{\partial \mathbf{H}} = \left(\mathbf{X}^*, \frac{\partial \mathbf{B}}{\partial \mathbf{H}} \right) = \int\limits_{t_0}^{t_1} \mathbf{X}^{*T}(t) \frac{\partial \mathbf{B}}{\partial \mathbf{H}} \, dt \qquad (6.128)$$

This expression has to be evaluated for each specific set of observables. In the case considered above, this set consists of the distance D and radial velocity V measured (simulated) at the end t_1 of the integration interval $[t_0, t_1]$.

In practical situations one has to deal with observables measured (simulated) at numerous instants t_R within the interval $[t_0, t_1]$. Use of a special adjoint solution, the adjoint propagator, provides an efficient way to deal with such situations. For the problem considered in this Section, the adjoint propagator $\mathbf{P}(t)$ is a 6×6 matrix solution of the finite-value problem:

$$\begin{cases} -\dfrac{d\mathbf{P}}{dt} + \mathbf{C}^T(t)\mathbf{P}(t) = \mathbf{O} \\ \mathbf{P}(t_1) = \mathbf{I} \end{cases} \qquad (6.129)$$

where the null-matrix $\mathbf{O}$ and identity matrix $\mathbf{I}$ are 6×6 matrices. Availability of this solution within the whole interval $[t_0, t_1]$ provides a way to compute the sensitivities for observables $\mathbf{R}$ measured (simulated) at any instant $t_R \in [t_0, t_1]$.

Indeed, assume that our observable is collected at some $t_R \in [t_0, t_1]$. Then, the corresponding adjoint problem has the form:

$$\begin{cases} -\dfrac{d\mathbf{X}_R^*}{dt} + \mathbf{C}^T(t)\mathbf{X}_R^*(t) = \mathbf{O} \\ \mathbf{X}_R^*(t_R) = \mathbf{W}_R \end{cases} \qquad (6.130)$$

Then, the solution $\mathbf{X}_R^*(t)$ of this adjoint problem can be expressed through the adjoint propagator as:

$$\mathbf{X}_R^*(t) = \mathbf{P}(t)\mathbf{P}_R^{-1}\mathbf{W}_R \qquad (6.131)$$

where $\mathbf{P}_R \equiv \mathbf{P}(t_R)$. Indeed, substituting Eq. (6.131) into the left side of the differential equation of the system Eq. (6.130), we have:

$$\left(-\frac{d}{dt} + \mathbf{C}^T(t) \right) \mathbf{P}(t) \mathbf{P}_R^{-1} \mathbf{W}_R = \mathbf{O} \qquad (6.132)$$

By taking the transpose of both sides of Eq. (6.132) we have:

$$\left(\mathbf{P}_R^{-1}\mathbf{W}_R \right)^T \left[\left(-\frac{d}{dt} + \mathbf{C}^T(t) \right) \mathbf{P}(t) \right]^T = \mathbf{O} \qquad (6.133)$$

Then the adjoint propagator $\mathbf{P}(t)$ satisfies the equation in the system Eq. (6.130)

$$\left(-\frac{d}{dt} + \mathbf{C}^T(t)\right)\mathbf{P}(t) = \mathbf{O} \tag{6.134}$$

Satisfaction of the final condition in Eq. (6.130) can be immediately verified. From Eq. (6.131) we have:

$$\mathbf{X}_R^*(t_R) = \mathbf{P}(t_R)\mathbf{P}_R^{-1}\mathbf{W}_R = \mathbf{P}_R\mathbf{P}_R^{-1}\mathbf{W}_R = \mathbf{W}_R \tag{6.135}$$

Thus, the matrix $\mathbf{X}_R^*(t)$ computed using the adjoint propagator $\mathbf{P}(t)$ is a solution of the adjoint problem Eq. (6.130) for any observable measured (simulated) anywhere within the integration interval $[t_0, t_1]$.

Note that the bulk of computations, which is associated with the adjoint propagator, do not depend on the specific choice of observables as long as they are drawn from the same forward solution $\mathbf{X}(t)$. This provides an additional flexibility in terms of choosing the observables, which are most informative for the particular inverse problems where the sensitivities of these observables will be used.

References

Sensitivity analysis of the model of radiances of a scattering planetary atmosphere in thermal spectral region was done in a number of the author's papers. The analysis in the form presented in this chapter was developed in (Ustinov 2001a) and (Ustinov 2002). The adjoint approach to sensitivity analysis of the zero-dimensional model of atmospheric dynamics was developed in (Ustinov 2001b) and is presented in this book in a slightly modified form. The linearized sensitivity analysis of the model of orbital tracking data of the planetary orbiter spacecraft is published in (Moyer 2000), and application of the adjoint approach to sensitivity analysis of this model was presented in (Ustinov and Sunseri 2005).

Moyer TD (2000) Formulation for observed and computed values of Deep Space Network data types for navigation. JPL Publication 00-7, JPL

Ustinov EA (2001a) Adjoint sensitivity analysis of radiative transfer equation: temperature and gas mixing ratio weighting functions for remote sensing of scattering atmospheres in thermal IR. J Quant Spectrosc Radiat Transf 68:195–211

Ustinov EA (2001b) Adjoint sensitivity analysis of atmospheric dynamics: application to the case of multiple observables. J Atmos Sci 58:3340–3348

Ustinov EA (2002) Adjoint sensitivity analysis of radiative transfer equation: 2. Applications to retrievals of temperature in scattering atmospheres in thermal IR. J Quant Spectrosc Radiat Transf 73:29–20

Ustinov EA, Sunseri RF (2005) Adjoint sensitivity analysis of orbital mechanics: application to computations of observables' partials with respect to harmonics of the planetary gravity fields. Paper presented at the General Assembly of European Geosciences Union, Vienna, Austria, 24–29 April 2005

Chapter 7
Sensitivity Analysis of Models with Higher-Order Differential Equations

Abstract All models considered in previous chapters are based on differential equations of first order. There exists a wide variety of models however, that are based on higher-order differential equations, such as the Poisson equation or wave equation. While application of the linearization approach to forward problems with these equations poses no substantial problems application of the adjoint approach, which needs formulation of corresponding adjoint problems, becomes more and more sophisticated with increasing the order of equations. In a nutshell, one has to apply the Lagrange identity rule as many times, as the order of the equations dictates, and this procedure becomes increasingly complicated [see, e.g., (Marchuk 1995)]. In this chapter we present an alternative approach based on the standard techniques using the reduction of the higher-order differential equation to a system of differential equations of first order. Further on, this system is represented in the form of a matrix differential equation of first order complemented by corresponding matrix initial-value conditions (IVCs) and/or boundary conditions (BCs). We present the general principles of application of this matrix approach and the results of its application to a set of problems based on selected stationary and non-stationary equations of mathematical physics.

Keywords Higher-order differential equations · Boundary conditions · Initial-value conditions · Final-value conditions · Indefinite conditions

7.1 General Principles of the Approach

In this section, we present the general principles of this approach for problems in 1D space. Applications to problems in 2D and 3D space will be considered in Sect. 7.4.

© The Author(s) 2015
E.A. Ustinov, *Sensitivity Analysis in Remote Sensing*,
SpringerBriefs in Earth Sciences, DOI 10.1007/978-3-319-15841-9_7

7.1.1 Stationary Problems

Following the standard technique mentioned above, we reduce the non-specified, higher-order differential equation to a system of 1st order equations. Let the scalar dependent variable u be the solution of this equation. Introducing a chain of additional dependent variables, each of which is a derivative of the previous one

$$\Gamma = u', \quad \Phi = \Gamma', \text{etc.} \tag{7.1}$$

and combining all dependent variables in a vector X, we transform the initial forward problem into a corresponding matrix forward problem. We first consider the case of a stationary differential equation with corresponding boundary conditions. Assuming that integration occurs on a 1D interval $[a, b]$, we have:

$$DX' + AX = S_e \tag{7.2}$$

$$B_a X|_a = S_a, \quad B_b X|_b = S_b \tag{7.3}$$

where D is a square matrix coefficient at the differential operator $(\ )'$; A, B_a, and B_b are square matrix coefficients; their rank corresponds to the dimension of the vector X; the right-hand terms S_e, S_a, and S_b are vectors of same dimension as X. The presence of *two* matrix BCs for a matrix differential equation of *first order* is justified because corresponding scalar BCs for additional variables contained in the vector X are *indefinite*, as will be demonstrated below.

To formulate the adjoint problemcorresponding to the forward problem, Eqs. (7.2) and (7.3) we convert this problem into a single operator equation in the form of Eq. (3.18). To ensure that BCs are acting only at the ends of the interval of integration we multiply them by delta-functions $\delta(x - a)$ and $\delta(x - b)$. Also, we multiply the BCs by factors c_a and c_b, which will be specified later. The resulting operator equation takes the form:

$$DX' + AX + c_a\delta(x - a)B_a X + c_b\delta(x - b)B_b X$$
$$= S_e + c_a\delta(x - a)S_a + c_b\delta(x - b)S_b \tag{7.4}$$

Then, the left side of the Lagrange identity Eq. (3.20) can be written as

$$(X^*, LX) = \int\limits_a^b dx X^{*T}(DX' + AX) + c_a\left(X^{*T}B_a X\right)\big|_a + c_b\left(X^{*T}B_b X\right)\big|_b \tag{7.5}$$

Integrating the integral term by parts and using the commutative law for transpose matrices, Eq. (A.7), we have:

$$\int_a^b X^{*T}(DX' + AX)\,dx = \int_a^b dx \left[-(D^T X^*)' + A^T X^*\right]^T X$$
$$+ \left[(D^T X^*)^T X\right]\Big|_b - \left[(D^T X^*)^T X\right]\Big|_a \qquad (7.6)$$

The off-integral terms in Eq. (7.5) are transformed as follows:

$$c_a(X^{*T} B_a X)\Big|_a = c_a\left[(B_a^T X^*)^T X\right]\Big|_a, \quad c_b(X^{*T} B_b X)\Big|_b = c_b\left[(B_b^T X^*)^T X\right]\Big|_b \qquad (7.7)$$

Then, choosing the specific values of factors c_a and c_b as

$$c_a = 1, \quad c_b = -1 \qquad (7.8)$$

we can rewrite Eq. (7.5) as

$$(X^*, LX) = \int_a^b dx \left[-(D^T X^*)' + A^T X^*\right]^T X$$
$$+ \left\{\left[(B_a - D)^T X^*\right]^T X\right\}\Big|_a - \left\{\left[(B_b - D)^T X^*\right]^T X\right\}\Big|_b \qquad (7.9)$$

The right-hand side of Eq. (7.9) can be represented in the form of the right-hand side of the Lagrange identity, Eq. (3.20), if we let

$$L^* X^* = -(D^T X^*)' + A^T X^* + \delta(x - a)(B_a - D)^T X^* - \delta(x - b)(B_b - D)^T X^* \qquad (7.10)$$

Correspondingly, the right-hand term of the resulting adjoint operator equation, Eq. (3.21) takes the form:

$$W = W_e + \delta(x - a)W_a - \delta(x - b)W_b \qquad (7.11)$$

Thus, we can formulate the matrix adjoint problem corresponding to the matrix forward problem, Eqs. (7.2) and (7.3), as follows:

$$-(D^T X^*)' + A^T X^* = W_e \qquad (7.12)$$

$$(B_a - D)^T X^*\Big|_a = W_a, \quad (B_b - D)^T X^*\Big|_b = W_b \qquad (7.13)$$

The general definition of the observable, Eq. (3.19) takes the form:

$$R = (W, X) = (W_e, X) + W_a X|_a - W_b X|_b \tag{7.14}$$

Also, with the choice of values of c_a and c_b per Eq. (7.8), the right-hand term of the operator equation Eq. (3.18) as represented by Eq. (7.4) has the form:

$$S = S_e + \delta(x - a)S_a - \delta(x - b)S_b \tag{7.15}$$

and the alternative definition of the observable, Eq. (3.22), takes the form:

$$R = (X^*, S_e) + X^*|_a S_a - X^*|_b S_b \tag{7.16}$$

7.1.2 Non-stationary Problems

If the initial forward problem is non-stationary, then additional variables are included into the forward solution X, which are corresponding time derivatives. The resulting matrix forward problem has the form:

$$C\dot{X} + DX' + AX = S_e \tag{7.17}$$

$$B_a X|_a = S_a(t), \quad B_b X|_b = S_b(t) \tag{7.18}$$

$$CX|_{t=0} = S_0(x) \tag{7.19}$$

The matrix initial-value condition (IVC) Eq. (7.19) has to be formulated so that the constant matrix C is identical to the matrix C in the matrix ordinary differential equation (ODE) Eq. (7.17). This is necessary to ensure proper conversion of the IVC of the matrix forward problem, Eq. (7.19) into the final-value condition (FVC) of the corresponding matrix adjoint problem, as will be presented below.

We can write the corresponding operator equation in the form analogous to Eq. (7.4) with values of c_a and c_b specified by Eq. (7.8):

$$C\dot{X} + DX' + AX + \delta(x - a)B_a X - \delta(x - b)B_b X + \delta(t)CX$$
$$= S_e + \delta(x - a)S_a - \delta(x - b)S_b + \delta(t)S_0 \tag{7.20}$$

To apply the Lagrange identity we assume that integration over time is performed in the interval $[0, T]$, where T is the length of this time interval (not to be confused with the "transpose" sign). The left side of Eq. (7.20) can be written as follows:

$$(X^*, LX) = \int_0^T dt \int_a^b dx X^{*T} \left(C\dot{X} + DX' + AX \right)$$

$$+ \int_0^T dt \left[(X^{*T} B_a X)\big|_a - (X^{*T} B_b X)\big|_b \right] + \int_a^b dx (X^{*T} C X)\big|_{t=0} \tag{7.21}$$

Re-grouping the terms and changing the order of integration over x and t where necessary, Eq. (7.21) can be further re-written as

$$(X^*, LX) = \int_a^b dx \left[\int_0^T dt X^{*T} C\dot{X} + (X^{*T} C X)\big|_{t=0} \right]$$

$$+ \int_0^T dt \left[\int_a^b dx X^{*T} (DX' + AX) + (X^{*T} B_a X)\big|_a - (X^{*T} B_b X)\big|_b \right] \tag{7.22}$$

The expression in the square brackets in the 2nd integral term of Eq. (7.22) is identical to that in the right side of Eq. (7.5) with values of constants c_a and c_b as in Eq. (7.8). Substituting the result of its transformation, Eq. (7.9), and integrating by parts the integral over t in the square brackets in the 1st integral term of Eq. (7.22), we obtain:

$$(X^*, LX) = \int_a^b dx \left\{ \int_0^T dt \left[-(C^T X^*)^{T} \dot{X} \right] + \left[(C^T X^*)^T X \right]\big|_{t=T} \right\}$$

$$+ \int_0^T dt \left\{ \int_a^b dx \left[-(D^T X^*)' + A^T X^* \right]^T X + \left[((B_a - D)^T X^*)^T X \right]\big|_a \right.$$

$$\left. - \left[((B_b - D)^T X^*)^T X \right]\big|_b \right. \tag{7.23}$$

The right side of Eq. (7.23) can be represented in the form of the right side of the Lagrange identity, Eq. (3.20), if we let

$$L^* X^* = - (C^T X^*)^{\cdot} - (D^T X^*)' + A^T X^*$$

$$+ \delta(x - a)(B_a - D)^T X^* - \delta(x - b)(B_b - D)^T X^* + \delta(t - T) C^T X^* \tag{7.24}$$

Correspondingly, the right-hand term of the resulting adjoint operator equation Eq. (3.21) takes the form:

$$W = W_e + \delta(x - a)W_a - \delta(x - b)W_b + \delta(t - T)W_T \tag{7.25}$$

The resulting matrix adjoint problem obtains the form:

$$- (C^T \overset{\bullet}{X}{}^*) - (D^T X^*)' + A^T X^* = W_e \tag{7.26}$$

$$(B_a - D)^T X^*|_a = W_a, \quad (B_b - D)^T X^*|_b = W_b \tag{7.27}$$

$$C^T X^*|_{t=T} = W_T \tag{7.28}$$

The Eq. (7.28) is a FVC imposed on the solution X^*. The definition of the observable, Eq. (3.19) takes the form:

$$R = (W, X) = (W_e, X) + W_a X|_a - W_b X|_b + W_0 X|_{t=T} \tag{7.29}$$

and the alternative definition of the observable, Eq. (3.21) takes the form:

$$R = (X^*, S_e) + X^*|_a S_a - X^*|_b S_b + X^*|_{t=0} S_0 \tag{7.30}$$

7.2 Applications to Stationary Problems

In this section, the general results obtained above will be applied to several stationary problems in 1D space.

7.2.1 Poisson Equation

We consider a forward problem with the Poisson equation and Dirichlet BC, which in general has the form

$$\Delta u = f(\mathbf{r}) \tag{7.31}$$

$$u|_S = f_S(\mathbf{r}_S) \tag{7.32}$$

Here, the forward solution $u(\mathbf{r})$ and the right-hand term of the equation $f(\mathbf{r})$ are scalar functions specified in some domain V of coordinates $\mathbf{r}$, and the right-hand term of the BC, $f(\mathbf{r}_S)$, is specified on the boundary S of this domain. In the 1D case, this forward problem has the form of an ODE with BCs:

$$u'' = f(x) \tag{7.33}$$

$$u|_a = f_a, \quad u|_b = f_b \tag{7.34}$$

Here, the forward solution $u(x)$ and the right-hand term $f(x)$ are scalar functions specified in some interval $[a, b]$ and the right-hand terms of the boundary conditions are specified at the ends of this interval $x = a$, and $x = b$.

To reduce the ODE of 2nd order Eq. (7.33) to a system of ODEs of 1st order we introduce the additional dependent variable $\Gamma = u'$ (gradient of u). Then, Eq. (7.33) becomes a system of two ODEs:

$$u' - \Gamma = 0 \tag{7.35}$$

$$\Gamma' = f \tag{7.36}$$

The system Eqs. (7.35) and (7.36) is further combined into a single matrix ODE of 1st order in the form of Eq. (7.2), where X is a two-component vector

$$X = \begin{pmatrix} u \\ \Gamma \end{pmatrix} \tag{7.37}$$

and the 2×2 matrix coefficient A along with the 2—vector S_e have the form:

$$A = \begin{pmatrix} 0 & -1 \\ 0 & 0 \end{pmatrix}, \quad S_e = \begin{pmatrix} 0 \\ f \end{pmatrix} \tag{7.38}$$

The matrix ODE in the form of Eq. (7.2) needs to be complemented by BCs for the vector X. To ensure that the component Γ of the vector X does not constrain the forward solution u, we have to impose on Γ the *indefinite* BCs in the form:

$$0 \cdot \Gamma|_a = 0, \quad 0 \cdot \Gamma|_b = 0 \tag{7.39}$$

Resulting BCs for the vector X have the form of Eq. (7.3), where

$$B_a = B_b = \begin{pmatrix} 1 & 0 \\ 0 & 0 \end{pmatrix} \tag{7.40}$$

and we obtain the resulting matrix forward problem in the expanded form of Eqs. (7.2) and (7.3) as follows:

$$\begin{pmatrix} 1 & 0 \\ 0 & 1 \end{pmatrix} \begin{pmatrix} u' \\ \Gamma' \end{pmatrix} + \begin{pmatrix} 0 & -1 \\ 0 & 0 \end{pmatrix} \begin{pmatrix} u \\ \Gamma \end{pmatrix} = \begin{pmatrix} 0 \\ f \end{pmatrix} \tag{7.41}$$

$$\begin{pmatrix} 1 & 0 \\ 0 & 0 \end{pmatrix} \begin{pmatrix} u \\ \Gamma \end{pmatrix}\bigg|_a = \begin{pmatrix} f_a \\ 0 \end{pmatrix}, \quad \begin{pmatrix} 1 & 0 \\ 0 & 0 \end{pmatrix} \begin{pmatrix} u \\ \Gamma \end{pmatrix}\bigg|_b = \begin{pmatrix} f_b \\ 0 \end{pmatrix} \tag{7.42}$$

Note that the matrix D of Eq. (7.2) is here an identity matrix.

To formulate the corresponding adjoint problem we need to specify its right-hand term. We assume that the observable R here is a scalar and has a simplest form

$$R = \int_a^b p(x)u(x) \, \mathrm{d}x = (p, u) \tag{7.43}$$

where $p(x)$ is the scalar observables weighting function specified on the domain $[a, b]$. Then the observable R can be expressed through the vector forward solution X as

$$R = \int_a^b W^T(x)X(x) \, \mathrm{d}x = (W, X) \tag{7.44}$$

where the matrix observables weighting function W has the form of a column vector

$$W = \begin{pmatrix} p \\ 0 \end{pmatrix} \tag{7.45}$$

Comparing Eq. (7.44) with the general expression for the right-hand term W of the operator adjoint equation Eq. (7.11), we conclude that its components have the form:

$$W_e = \begin{pmatrix} p \\ 0 \end{pmatrix} = W, \quad W_a = \begin{pmatrix} 0 \\ 0 \end{pmatrix}, \quad W_b = \begin{pmatrix} 0 \\ 0 \end{pmatrix}, \tag{7.46}$$

Thus, the matrix adjoint problem Eqs. (7.12) and (7.13) in the expanded form is as follows:

$$-\begin{pmatrix} 1 & 0 \\ 0 & 1 \end{pmatrix}\begin{pmatrix} u^{*\prime} \\ \Gamma^{*\prime} \end{pmatrix} + \begin{pmatrix} 0 & 0 \\ -1 & 0 \end{pmatrix}\begin{pmatrix} u^{*} \\ \Gamma^{*} \end{pmatrix} = \begin{pmatrix} p \\ 0 \end{pmatrix} \tag{7.47}$$

$$\begin{pmatrix} 0 & 0 \\ 0 & -1 \end{pmatrix}\begin{pmatrix} u^{*} \\ \Gamma^{*} \end{pmatrix}\Big|_a = \begin{pmatrix} 0 \\ 0 \end{pmatrix}, \quad \begin{pmatrix} 0 & 0 \\ 0 & -1 \end{pmatrix}\begin{pmatrix} u^{*} \\ \Gamma^{*} \end{pmatrix}\Big|_b = \begin{pmatrix} 0 \\ 0 \end{pmatrix} \tag{7.48}$$

and the adjoint vector solution X^* is a two-component vector

$$X^* = \begin{pmatrix} u^{*} \\ \Gamma^{*} \end{pmatrix} \tag{7.49}$$

Rewriting the matrix adjoint problem in the form of a system of scalar ODEs and BCs, we have:

$$-u^{*\prime} = p \tag{7.50}$$

$$-\Gamma^{*\prime} - u^* = 0 \tag{7.51}$$

$$0 \cdot u^*|_a = 0, \quad 0 \cdot u^*|_b = 0 \tag{7.52}$$

$$\Gamma^*|_a = 0, \quad \Gamma^*|_b = 0 \tag{7.53}$$

Note that here we have indefinite BCs, Eq. (7.52), for the component u^*, which will be discarded.

Now we convert the matrix adjoint problem Eqs. (7.47) and (7.48) to a set of scalar ODEs with scalar BCs. From Eq. (7.51) we have:

$$-u^{*\prime} = \Gamma^{*\prime\prime} \tag{7.54}$$

Substituting in Eq. (7.50) and complementing the resulting ODE of 2nd order with corresponding BCs, Eq. (7.53), we have:

$$\Gamma^{*\prime\prime} = p \tag{7.55}$$

$$\Gamma^*|_a = 0, \quad \Gamma^*|_b = 0 \tag{7.56}$$

Thus, the component Γ^* of the adjoint vector solution X^* can be interpreted as the sought for adjoint scalar solution:

$$\Gamma^* = w \tag{7.57}$$

Substituting Eq. (7.57) in Eqs. (7.55) and (7.56), we obtain the adjoint problem corresponding to the forward problem, Eqs. (7.33) and (7.34):

$$w'' = p \tag{7.58}$$

$$w|_a = 0, \quad w|_b = 0 \tag{7.59}$$

Thus, we have derived the scalar adjoint problem for the given scalar forward problem Eqs. (7.33), (7.34). To do that, we have:

1. converted the initial scalar forward problem to a matrix form, corresponding to Eqs. (7.2) and (7.3),
2. formulated the corresponding matrix adjoint problem using its general form, Eqs. (7.12) and (7.13), and
3. converted the matrix adjoint problem to a scalar adjoint problem.

It should be emphasized that these derivations were based on the results of the previous section, which were obtained in the general matrix-vector form, whereas the specifics of the ODE and BCs of the initial forward problem were encapsulated in corresponding matrices and vectors. This approach will be used throughout this chapter for various cases of stationary and non-stationary problems.

Now we consider the Poisson equation Eq. (7.33) with Neumann BCs, which specify the boundary values of the derivative u' of the forward solution, rather than the boundary values of the forward solution u itself:

$$u'|_a = f_a, \quad u'|_b = f_a \tag{7.60}$$

Combining with indefinite BCs imposed on the forward solution u

$$0 \cdot u|_a = 0, \quad 0 \cdot u|_b = 0 \tag{7.61}$$

we obtain the corresponding matrix forward problem in the form:

$$\begin{pmatrix} 1 & 0 \\ 0 & 1 \end{pmatrix} \begin{pmatrix} u' \\ \Gamma' \end{pmatrix} + \begin{pmatrix} 0 & -1 \\ 0 & 0 \end{pmatrix} \begin{pmatrix} u \\ \Gamma \end{pmatrix} = \begin{pmatrix} 0 \\ f \end{pmatrix} \tag{7.62}$$

$$\begin{pmatrix} 0 & 0 \\ 0 & 1 \end{pmatrix} \begin{pmatrix} u \\ \Gamma \end{pmatrix}\bigg|_a = \begin{pmatrix} 0 \\ f_a \end{pmatrix}, \quad \begin{pmatrix} 0 & 0 \\ 0 & 1 \end{pmatrix} \begin{pmatrix} u \\ \Gamma \end{pmatrix}\bigg|_b = \begin{pmatrix} 0 \\ f_b \end{pmatrix} \tag{7.63}$$

Using the general results of the previous section, we can immediately formulate the corresponding matrix adjoint problem in the expanded form:

$$-\begin{pmatrix} 1 & 0 \\ 0 & 1 \end{pmatrix} \begin{pmatrix} u*' \\ \Gamma^{*'} \end{pmatrix} + \begin{pmatrix} 0 & 0 \\ -1 & 0 \end{pmatrix} \begin{pmatrix} u^* \\ \Gamma^* \end{pmatrix} = \begin{pmatrix} p \\ 0 \end{pmatrix} \tag{7.64}$$

$$\begin{pmatrix} -1 & 0 \\ 0 & 0 \end{pmatrix} \begin{pmatrix} u^* \\ \Gamma^* \end{pmatrix}\bigg|_a = \begin{pmatrix} 0 \\ 0 \end{pmatrix}, \quad \begin{pmatrix} -1 & 0 \\ 0 & 0 \end{pmatrix} \begin{pmatrix} u^* \\ \Gamma^* \end{pmatrix}\bigg|_b = \begin{pmatrix} 0 \\ 0 \end{pmatrix} \tag{7.65}$$

Rewriting Eqs. (7.64) and (7.65) in the form of a system of scalar ODEs and BCs, we have:

$$-u^{*'} = p \tag{7.66}$$

$$-\Gamma^{*'} - u^* = 0 \tag{7.67}$$

$$u^*|_a = 0, \quad u^*|_b = 0 \tag{7.68}$$

$$0 \cdot \Gamma^*|_a = 0, \quad 0 \cdot \Gamma^*|_b = 0 \tag{7.69}$$

Acting in a fashion similar to the previous case, with Dirichlet BCs, and taking into account Eq. (7.67), we have:

$$\Gamma^{*\prime\prime} = p \tag{7.70}$$

$$\Gamma^{*\prime}|_a = 0, \quad \Gamma^{*\prime}|_b = 0 \tag{7.71}$$

The component Γ^* of the adjoint vector solution X^* can, again, be interpreted as the sought for adjoint scalar solution Eq. (7.57). Substituting Eq. (7.57) in Eqs. (7.70) and (7.71), we obtain the resulting adjoint problem corresponding to the forward problem, Eqs. (7.33) and (7.60):

$$w^{\prime\prime} = p \tag{7.72}$$

$$w^{\prime}|_a = 0, \quad w^{\prime}|_b = 0 \tag{7.73}$$

Thus, the scalar adjoint problem, Eqs. (7.72) and (7.73) has a form of the Poisson equation with Neumann boundary conditions, again, as in the previous case, essentially identical to the initial scalar forward problem.

7.2.2 Bi-harmonic Equation

The forward problem consisting of the biharmonic equation with Dirichlet BCs of 2nd kind has the form:

$$\Delta^2 u = f(\mathbf{r}) \tag{7.74}$$

$$u|_S = f_S^{(0)}(\mathbf{r}_S) \tag{7.75}$$

$$\Delta u|_S = f_S^{(2)}(\mathbf{r}_S) \tag{7.76}$$

In 1D case, this forward problem has a form of an ODE of 4th order with BCs:

$$u^{IV} = f(x) \tag{7.77}$$

$$u|_a = f_a^{(0)}, \quad u|_b = f_b^{(0)} \tag{7.78}$$

$$u^{\prime\prime}|_a = f_a^{(2)}, \quad u^{\prime\prime}|_b = f_b^{(2)} \tag{7.79}$$

To reduce the ODE Eq. (7.77) to a system of ODEs of 1st order we introduce the additional dependent variables $\Gamma = u', \Phi = \Gamma', $ and $\Psi = \Phi'$. Then, we obtain a system of four ODEs:

$$u' - \Gamma = 0 \tag{7.80}$$

$$\Gamma' - \Phi = 0 \tag{7.81}$$

$$\Phi' - \Psi = 0 \tag{7.82}$$

$$\Psi' = f \tag{7.83}$$

The BCs of 2nd order, Eq. (7.79), are rewritten as

$$\Phi|_a = f_a^{(2)}, \quad \Phi|_b = f_b^{(2)} \tag{7.84}$$

To ensure that the components Γ and Ψ do not constrain the forward solution u, we impose two pairs of indefinite BCs in the form:

$$0 \cdot \Gamma|_a = 0, \quad 0 \cdot \Gamma|_b = 0; \quad 0 \cdot \Psi|_a = 0, \quad 0 \cdot \Psi|_b = 0 \tag{7.85}$$

The resulting matrix forward problem in the expanded form of Eqs. (7.2) and (7.3) is as follows:

$$\begin{pmatrix} 1 & 0 & 0 & 0 \\ 0 & 1 & 0 & 0 \\ 0 & 0 & 1 & 0 \\ 0 & 0 & 0 & 1 \end{pmatrix} \begin{pmatrix} u' \\ \Gamma' \\ \Phi' \\ \Psi' \end{pmatrix} + \begin{pmatrix} 0 & -1 & 0 & 0 \\ 0 & 0 & -1 & 0 \\ 0 & 0 & 0 & -1 \\ 0 & 0 & 0 & 0 \end{pmatrix} \begin{pmatrix} u \\ \Gamma \\ \Phi \\ \Psi \end{pmatrix} = \begin{pmatrix} 0 \\ 0 \\ 0 \\ f \end{pmatrix} \tag{7.86}$$

$$\begin{pmatrix} 1 & 0 & 0 & 0 \\ 0 & 0 & 0 & 0 \\ 0 & 0 & 1 & 0 \\ 0 & 0 & 0 & 0 \end{pmatrix} \begin{pmatrix} u \\ \Gamma \\ \Phi \\ \Psi \end{pmatrix} = \begin{pmatrix} f_a^{(0)} \\ 0 \\ f_a^{(2)} \\ 0 \end{pmatrix}, \quad \begin{pmatrix} 1 & 0 & 0 & 0 \\ 0 & 0 & 0 & 0 \\ 0 & 0 & 1 & 0 \\ 0 & 0 & 0 & 0 \end{pmatrix} \begin{pmatrix} u \\ \Gamma \\ \Phi \\ \Psi \end{pmatrix} = \begin{pmatrix} f_b^{(0)} \\ 0 \\ f_b^{(2)} \\ 0 \end{pmatrix} \tag{7.87}$$

The corresponding matrix adjoint problem has the form of Eqs. (7.12) and (7.13). To formulate its right-hand term, we assume the observable R in the form of Eq. (7.43). Then, from comparison with the general expression Eq. (7.44), we conclude that the matrix observables weighting function W has here the form of a column vector

$$W = \begin{pmatrix} p \\ 0 \\ 0 \\ 0 \end{pmatrix} \tag{7.88}$$

Repeating the pattern of derivations of the previous Subsection, we obtain the sought for matrix adjoint problem in the expanded form:

$$-\begin{pmatrix} 1 & 0 & 0 & 0 \\ 0 & 1 & 0 & 0 \\ 0 & 0 & 1 & 0 \\ 0 & 0 & 0 & 1 \end{pmatrix} \begin{pmatrix} u*' \\ \Gamma^{*\prime} \\ \Phi^{*\prime} \\ \Psi^{*\prime} \end{pmatrix} + \begin{pmatrix} 0 & 0 & 0 & 0 \\ -1 & 0 & 0 & 0 \\ 0 & -1 & 0 & 0 \\ 0 & 0 & -1 & 0 \end{pmatrix} \begin{pmatrix} u^* \\ \Gamma^* \\ \Phi^* \\ \Psi^* \end{pmatrix} = \begin{pmatrix} p \\ 0 \\ 0 \\ 0 \end{pmatrix} \quad (7.89)$$

$$\begin{pmatrix} 0 & 0 & 0 & 0 \\ 0 & -1 & 0 & 0 \\ 0 & 0 & 0 & 0 \\ 0 & 0 & 0 & -1 \end{pmatrix} \begin{pmatrix} u^* \\ \Gamma^* \\ \Phi^* \\ \Psi^* \end{pmatrix} = \begin{pmatrix} 0 \\ 0 \\ 0 \\ 0 \end{pmatrix}, \quad \begin{pmatrix} 0 & 0 & 0 & 0 \\ 0 & -1 & 0 & 0 \\ 0 & 0 & 0 & 0 \\ 0 & 0 & 0 & -1 \end{pmatrix} \begin{pmatrix} u^* \\ \Gamma^* \\ \Phi^* \\ \Psi^* \end{pmatrix} = \begin{pmatrix} 0 \\ 0 \\ 0 \\ 0 \end{pmatrix}$$

$$(7.90)$$

Now we rewrite Eqs. (7.89) and (7.90) in the form of a system of scalar ODEs and BCs. After discarding the indefinite scalar BCs for u^* and Φ^*, we have:

$$-u^{*\prime} = p \quad (7.91)$$

$$-\Gamma^{*\prime} - u^* = 0 \quad (7.92)$$

$$-\Phi^{*\prime} - \Gamma^* = 0 \quad (7.93)$$

$$-\Psi^{*\prime} - \Phi^* = 0 \quad (7.94)$$

$$\Gamma^*|_a = 0, \quad \Gamma^*|_b = 0 \quad (7.95)$$

$$\Psi^*|_a = 0, \quad \Psi^*|_b = 0 \quad (7.96)$$

From Eqs. (7.92)–(7.94) we have:

$$-u^{*\prime} = \Gamma^{*\prime\prime} = -\Phi^{*\prime\prime\prime} = \Psi^{*IV} \quad (7.97)$$

Substituting equalities Eq. (7.97) in Eq. (7.91), we obtain the scalar adjoint differential equation of 4th order:

$$\Psi^{*IV} = p \quad (7.98)$$

On the other hand, from Eqs. (7.93) and (7.94), we have:

$$\Gamma^* = -\Phi^{*\prime} = \Psi^{*\prime\prime} \quad (7.99)$$

Therefore, the first BC, Eq. (7.95) can be written as

$$\Psi^{*\prime\prime}|_a = 0, \quad \Psi^{*\prime\prime}|_b = 0 \quad (7.100)$$

The component Ψ of the adjoint vector solution X^* can, therefore, be interpreted as the sought for adjoint scalar solution

$$w = \Psi^* \tag{7.101}$$

Substituting Eq. (7.101) in Eqs. (7.92) and (7.96), we obtain the resulting adjoint problem corresponding to the forward problem, Eqs. (7.77)–(7.79):

$$w^{IV} = p \tag{7.102}$$

$$w|_a = 0, \quad w|_b = 0 \tag{7.103}$$

$$w''|_a = 0, \quad w|_b = 0 \tag{7.104}$$

The resulting scalar adjoint problem, Eqs. (7.102)–(7.104), has a form of the bi-harmonic equation with Dirichlet boundary conditions of 2nd kind, again as in the previous case, essentially identical to the initial scalar forward problem.

This concludes consideration of stationary 1D forward problems with higher-order ODEs. In the next section we consider formulation of adjoint problems for non-stationary 1D forward problems.

7.3 Applications to Non-stationary Problems

7.3.1 Heat Equation

The heat equation with Dirichlet BCs and a corresponding IVC has the form:

$$\frac{\partial u}{\partial t} - \alpha(\mathbf{r}, t)\Delta u = f(\mathbf{r}, t) \tag{7.105}$$

$$u|_S = f_S(\mathbf{r}_S, t) \tag{7.106}$$

$$u|_{t=0} = f_0(\mathbf{r}) \tag{7.107}$$

In the 1D case, we write the forward problem, Eqs. (7.105)–(7.107) in the form:

$$-\dot{u} + \alpha(x, t)u'' = -f(x, t) \tag{7.108}$$

$$u|_a = f_a(t), \quad u|_b = f(t)_b \tag{7.109}$$

$$u|_{t=0} = f_0(x) \tag{7.110}$$

First, we consider the case when $\alpha(x, t) = $ const. Then, by the same substitution $u' = \Gamma$, as before, we reduce the equation Eq. (7.108) to a system of two equations

$$u' - \Gamma = 0 \tag{7.111}$$

$$-\overset{\bullet}{u} + \alpha\Gamma' = -f \tag{7.112}$$

For the function Γ we impose indefinite BCs in the form of Eq. (7.39) and an indefinite IVC in the form:

$$0 \cdot \Gamma|_{t=0} = 0 \tag{7.113}$$

The resulting matrix forward problem in the expanded form can be written as:

$$\begin{pmatrix} 0 & 0 \\ -1 & 0 \end{pmatrix} \begin{pmatrix} \overset{\bullet}{u} \\ \overset{\bullet}{\Gamma} \end{pmatrix} + \begin{pmatrix} 1 & 0 \\ 0 & \alpha \end{pmatrix} \begin{pmatrix} u' \\ \Gamma' \end{pmatrix} + \begin{pmatrix} 0 & -1 \\ 0 & 0 \end{pmatrix} \begin{pmatrix} u \\ \Gamma \end{pmatrix} = \begin{pmatrix} 0 \\ -f \end{pmatrix} \tag{7.114}$$

$$\begin{pmatrix} 1 & 0 \\ 0 & 0 \end{pmatrix} \begin{pmatrix} u \\ \Gamma \end{pmatrix}\bigg|_a = \begin{pmatrix} f_a \\ 0 \end{pmatrix}, \quad \begin{pmatrix} 1 & 0 \\ 0 & 0 \end{pmatrix} \begin{pmatrix} u \\ \Gamma \end{pmatrix}\bigg|_b = \begin{pmatrix} f_b \\ 0 \end{pmatrix} \tag{7.115}$$

$$\begin{pmatrix} 0 & 0 \\ -1 & 0 \end{pmatrix} \begin{pmatrix} u \\ \Gamma \end{pmatrix}\bigg|_{t=0} = \begin{pmatrix} 0 \\ -f_0 \end{pmatrix} \tag{7.116}$$

This matrix forward problem has the form of Eqs. (7.17)–(7.19), and corresponding matrix adjoint problem has the form of Eqs. (7.26)–(7.28). In the expanded form we have:

$$-\begin{pmatrix} 0 & -1 \\ 0 & 0 \end{pmatrix} \begin{pmatrix} \overset{\bullet}{u^*} \\ \overset{\bullet}{\Gamma^*} \end{pmatrix} - \left[\begin{pmatrix} 1 & 0 \\ 0 & \alpha \end{pmatrix} \begin{pmatrix} u^* \\ \Gamma^* \end{pmatrix} \right]' + \begin{pmatrix} 0 & 0 \\ -1 & 0 \end{pmatrix} \begin{pmatrix} u^* \\ \Gamma^* \end{pmatrix} = \begin{pmatrix} p \\ 0 \end{pmatrix} \tag{7.117}$$

$$\begin{pmatrix} 0 & 0 \\ 0 & -\alpha \end{pmatrix} \begin{pmatrix} u* \\ \Gamma^* \end{pmatrix}\bigg|_a = \begin{pmatrix} 0 \\ 0 \end{pmatrix}, \quad \begin{pmatrix} 0 & 0 \\ 0 & -\alpha \end{pmatrix} \begin{pmatrix} u^* \\ \Gamma^* \end{pmatrix}\bigg|_b = \begin{pmatrix} 0 \\ 0 \end{pmatrix} \tag{7.118}$$

$$\begin{pmatrix} 0 & -1 \\ 0 & 0 \end{pmatrix} \begin{pmatrix} u* \\ \Gamma^* \end{pmatrix}\bigg|_{t=T} = \begin{pmatrix} 0 \\ 0 \end{pmatrix} \tag{7.119}$$

Now we rewrite Eqs. (7.117)–(7.119) in the form of a system of scalar ODEs with corresponding BCs and an IVC. After discarding the indefinite BCs and FVC at $t = T$, we have:

$$\overset{\bullet}{\Gamma^*} - u^{*\prime} = p \tag{7.120}$$

$$-(\alpha\Gamma^*)' - u^* = 0 \tag{7.121}$$

$$\Gamma^*|_a = 0, \ \Gamma^*|_b = 0 \qquad\qquad (7.122)$$

$$\Gamma^*|_{t=T} = 0 \qquad\qquad (7.123)$$

From Eq. (7.121) we have:

$$-u^{*\prime} = (\alpha\Gamma^{*\prime})' \qquad\qquad (7.124)$$

Substituting in Eq. (7.120), we obtain:

$$\dot{\Gamma}^* + (\alpha\Gamma^{*\prime})' = p \qquad\qquad (7.125)$$

Introducing the scalar adjoint solution $w = \Gamma^*$, Eq. (7.57), we rewrite the ODE Eq. (7.125), BCs Eq. (7.122) and FVC Eq. (7.123) in the form of the resulting adjoint problem corresponding to the initial forward problem, Eqs. (7.108)–(7.110):

$$\dot{w} + (\alpha w')' = p \qquad\qquad (7.126)$$

$$w|_a = 0, \quad w|_b = 0 \qquad\qquad (7.127)$$

$$w|_{t=T} = 0 \qquad\qquad (7.128)$$

The above derivation was based on the assumption that α is a single coefficient. In general, this is not case, and the heat equation has the form:

$$\rho c \frac{\partial u}{\partial t} - \nabla^T(k\nabla u) = f \qquad\qquad (7.129)$$

where ρ is the mass density, and c and k are coefficients of heat capacity and thermal conductivity, which may depend on the space coordinates. (For the sake of simplicity, we make a realistic assumption that ρ and c do not depend on time.) In the 1D case, Eq. (7.129) can be written in the form:

$$-\rho c\,\dot{u} + (ku')' = -f \qquad\qquad (7.130)$$

Here we introduce the intermediate variable $\Gamma = ku'$ so that

$$u' - \frac{\Gamma}{k} = 0 \qquad\qquad (7.131)$$

Substituting Eq. (7.131) in ODE Eq. (7.130), we have:

$$-\rho c\,\dot{u} + \Gamma' = -f \qquad\qquad (7.132)$$

Combining Eqs. (7.131) and (7.132) into a matrix ODE, we have:

$$\begin{pmatrix} 0 & 0 \\ -\rho c & 0 \end{pmatrix} \begin{pmatrix} \dot{u} \\ \dot{\Gamma} \end{pmatrix} + \begin{pmatrix} 1 & 0 \\ 0 & 1 \end{pmatrix} \begin{pmatrix} u' \\ \Gamma' \end{pmatrix} + \begin{pmatrix} 0 & -1/k \\ 0 & 0 \end{pmatrix} \begin{pmatrix} u \\ \Gamma \end{pmatrix} = \begin{pmatrix} 0 \\ -f \end{pmatrix} \quad (7.133)$$

We will use the Dirichlet BCs, Eq. (7.109). As for the IVC, we multiply the scalar IVC, Eq. (7.110) by a factor $-\rho c$ to make the matrix C of the resulting matrix IVC

$$\begin{pmatrix} 0 & 0 \\ -\rho c & 0 \end{pmatrix} \begin{pmatrix} u \\ \Gamma \end{pmatrix} \bigg|_{t=0} = \begin{pmatrix} 0 \\ -\rho c f_0 \end{pmatrix} \quad (7.134)$$

to be identical to the matrix C of the matrix ODE Eq. (7.133).

Now we can directly apply the general results of Sect. 7.1. The corresponding matrix adjoint problem in the expanded form is written as:

$$-\begin{pmatrix} 0 & -\rho c \\ 0 & 0 \end{pmatrix} \begin{pmatrix} \dot{u}^* \\ \dot{\Gamma}^* \end{pmatrix} - \begin{pmatrix} 1 & 0 \\ 0 & 1 \end{pmatrix} \begin{pmatrix} u^{*'} \\ \Gamma^{*'} \end{pmatrix} + \begin{pmatrix} 0 & 0 \\ -1/k & 0 \end{pmatrix} \begin{pmatrix} u^* \\ \Gamma^* \end{pmatrix} = \begin{pmatrix} p \\ 0 \end{pmatrix}$$

$$\quad (7.135)$$

$$\begin{pmatrix} 0 & 0 \\ 0 & -1 \end{pmatrix} \begin{pmatrix} u^* \\ \Gamma^* \end{pmatrix} \bigg|_a = \begin{pmatrix} 0 \\ 0 \end{pmatrix}, \quad \begin{pmatrix} 0 & 0 \\ 0 & -1 \end{pmatrix} \begin{pmatrix} u \\ \Gamma \end{pmatrix} \bigg|_b = \begin{pmatrix} 0 \\ 0 \end{pmatrix} \quad (7.136)$$

$$\begin{pmatrix} 0 & -\rho c \\ 0 & 0 \end{pmatrix} \begin{pmatrix} u^* \\ \Gamma^* \end{pmatrix} \bigg|_{t=T} = \begin{pmatrix} 0 \\ 0 \end{pmatrix} \quad (7.137)$$

Rewriting Eqs. (7.135)–(7.137) in the form of a system of scalar ODEs with the corresponding BCs and FVC, and discarding the indefinite BCs and FVC, we have:

$$\rho c \, \dot{\Gamma}^* - u^{*'} = p \quad (7.138)$$

$$-\Gamma^{*'} - \frac{1}{k} u^* = 0 \quad (7.139)$$

$$\Gamma^*|_a = 0, \quad \Gamma^*|_b = 0 \quad (7.140)$$

$$\Gamma^*|_{t=T} = 0 \quad (7.141)$$

From Eq. (7.139) we have:

$$-u^{*'} = (k\Gamma^{*'})' \quad (7.142)$$

Substituting in Eq. (7.138), we obtain:

$$\rho c \, \dot{\Gamma}^* + (k\Gamma^{*\prime})' = p \qquad (7.143)$$

And, introducing the scalar adjoint solution $w = \Gamma^*$, we rewrite the PDE Eq. (7.143), BCs Eq. (7.140) and FVC Eq. (7.141) in the form of the scalar adjoint problem corresponding to the forward problem, Eqs. (7.130), (7.109) and (7.110):

$$\rho c \, \dot{w} + (kw')' = p \qquad (7.144)$$

$$w|_a = 0, \quad w|_b = 0 \qquad (7.145)$$

$$w|_{t=T} = 0 \qquad (7.146)$$

Note that in this derivation we have implicitly assumed that the product ρc does not depend on time.

With the adjoint solution $w(x)$ at hand, we can use the general expression, Eq. (3.26), to obtain the sensitivity $\delta R / \delta k(x)$. Replacing the non-linear operator N in Eq. (3.26) by the linear operator L of the problem under consideration and observing that $\delta S / \delta k(x) = 0$, we have:

$$\frac{\delta R}{\delta k(x)} = \left(w, \frac{\delta L}{\delta k(x)} u \right) \qquad (7.147)$$

Note that the parameter $k(x)$ is situated in the operator L under the sign of derivative. This complication can be resolved in a following way. Observing that

$$\left(w, \frac{\delta L}{\delta k(x)} u \right) \equiv \left(w, \frac{\delta[(ku')']}{\delta k(x)} \right) \qquad (7.148)$$

and re-writing the corresponding term of the operator L as

$$(ku')' = ku'' + k'u' \qquad (7.149)$$

we have:

$$\left(w, \frac{\delta[(ku')']}{\delta k(x)} \right) = \int\limits_0^T \mathrm{d}t \int\limits_a^b \mathrm{d}\xi \, w(\xi, t) \left\{ \frac{\delta[k(\xi)]}{\delta k(x)} u''(\xi, t) + \frac{\delta[k'(\xi)]}{\delta k(x)} u'(\xi, t) \right\} \qquad (7.150)$$

For the first intra-integral term in Eq. (7.150) we have [cf. Eq. (2.11)]:

$$\int\limits_0^T dt \int\limits_a^b d\xi\, w(\xi,t) \frac{\delta[k(\xi)]}{\delta k(x)} u''(\xi,t) = \int\limits_0^T dt \int\limits_a^b d\xi\, w(\xi,t)\delta(\xi - x)u''(\xi,t)$$

$$= \int\limits_0^T w(x,t)u''(x,t)\, dt \qquad (7.151)$$

For the second intra-integral term, we have [cf. Eq. (2.12)]:

$$\int\limits_0^T dt \int\limits_a^b d\xi\, w(\xi,t) \frac{\delta[k'(\xi)]}{\delta k(x)} u'(\xi,t) = -\int\limits_0^T dt \int\limits_a^b d\xi\, w(\xi,t)\delta'(\xi - x)u'(\xi,t)$$

$$(7.152)$$

Integrating the integral over the interval $[a, b]$ by parts:

$$\int\limits_a^b d\xi\, w(\xi,t)\delta'(\xi - x)u'(\xi,t) = \delta(\xi - x)w(\xi,t)u'(\xi,t)\big|_a^b$$

$$- \int\limits_a^b d\xi\, \delta(\xi - x)[w(\xi,t)u'(\xi,t)]'$$

$$= \delta(\xi - x)w(\xi,t)u'(\xi,t)\big|_a^b - [w(x,t)u'(x,t)]' \qquad (7.153)$$

and substituting the result in Eq. (7.152), we obtain that everywhere within the interval $[a < x < b]$:

$$\int\limits_0^T dt \int\limits_a^b d\xi\, w(\xi,t) \frac{\delta[k'(\xi)]}{\delta k(x)} u'(\xi,t) = \int\limits_0^T [w(x,t)u'(x,t)]'\, dt \qquad (7.154)$$

Then, substituting Eqs. (7.151) and (7.154) into Eq. (7.150) and the result in Eq. (7.148) and then in Eq. (7.147), we obtain:

$$\frac{\delta R}{\delta k(x)} = \int\limits_0^T \left\{ w(x,t)u''(x,t) + [w(x,t)u'(x,t)]' \right\} dt \qquad (7.155)$$

The baseline forward solution $u(x, t)$ is available in a numerical form. Therefore, its derivatives $u'(x, t)$ and $u''(x, t)$ are also available via suitable numerical differentiation schemes.

7.3.2 Wave Equation

The wave equation with Dirichlet BCs and a corresponding IVC has the form:

$$-\frac{1}{c^2}\frac{\partial^2 u}{\partial t^2} + \Delta u = f(\mathbf{r}, t) \tag{7.156}$$

$$u|_S = f_S(\mathbf{r}_S, t) \tag{7.157}$$

$$u|_{t=0} = f_0(\mathbf{r}), \quad \dot{u}\Big|_{t=0} = f_1(\mathbf{r}) \tag{7.158}$$

where c is the propagation speed of the wave. In the 1D case, denoting $\alpha \equiv 1/c^2$ we write the forward problem, Eqs. (7.156)–(7.158) in the form:

$$-\alpha(x, t)\,\ddot{u} + u'' = f(x, t) \tag{7.159}$$

$$u|_a = f_a(t), \quad u|_b = f(t)_b \tag{7.160}$$

$$u|_{t=0} = f_0(x), \quad \dot{u}\Big|_{t=0} = f_1(x) \tag{7.161}$$

Here we introduce the intermediate variables $v = \dot{u}$ and $\Gamma = u'$, so that

$$\dot{u} - v = 0 \tag{7.162}$$

and

$$u' - \Gamma = 0 \tag{7.163}$$

Then, Eq. (7.159) can be rewritten in the form:

$$-\alpha\dot{v} + \Gamma' = f \tag{7.164}$$

Combining Eqs. (7.162)–(7.164) into a matrix ODE, we have:

$$\begin{pmatrix} 0 & 1 & 0 \\ 0 & 0 & 0 \\ -\alpha & 0 & 0 \end{pmatrix}\begin{pmatrix} \dot{v} \\ \dot{u} \\ \dot{\Gamma} \end{pmatrix} + \begin{pmatrix} 0 & 0 & 0 \\ 0 & 1 & 0 \\ 0 & 0 & 1 \end{pmatrix}\begin{pmatrix} v' \\ u' \\ \Gamma' \end{pmatrix} + \begin{pmatrix} -1 & 0 & 0 \\ 0 & 0 & -1 \\ 0 & 0 & 0 \end{pmatrix}\begin{pmatrix} v \\ u \\ \Gamma \end{pmatrix} = \begin{pmatrix} 0 \\ 0 \\ f \end{pmatrix} \tag{7.165}$$

Complementing the Dirichlet BCs, Eq. (7.160) by two indefinite BCs for components v and Γ, we obtain the matrix BCs in the form:

$$\begin{pmatrix} 0 & 1 & 0 \\ 0 & 0 & 0 \\ 0 & 0 & 0 \end{pmatrix} \begin{pmatrix} v \\ u \\ \Gamma \end{pmatrix}\bigg|_a = \begin{pmatrix} f_a \\ 0 \\ 0 \end{pmatrix}, \quad \begin{pmatrix} 0 & 1 & 0 \\ 0 & 0 & 0 \\ 0 & 0 & 0 \end{pmatrix} \begin{pmatrix} v \\ u \\ \Gamma \end{pmatrix}\bigg|_b = \begin{pmatrix} f_b \\ 0 \\ 0 \end{pmatrix} \quad (7.166)$$

Similarly, complementing the IVC Eq. (7.161) by an indefinite IVC for the component Γ we obtain the matrix IVC in the form:

$$\begin{pmatrix} 0 & 1 & 0 \\ 1 & 0 & 0 \\ 0 & 0 & 0 \end{pmatrix} \begin{pmatrix} v \\ u \\ \Gamma \end{pmatrix}\bigg|_{t=0} = \begin{pmatrix} f_0 \\ f_1 \\ 0 \end{pmatrix} \quad (7.167)$$

To proceed, we need to interchange the upper two rows of the matrix coefficient B and right-hand term in the matrix BCs, Eq. (7.166), to make compatible the matrix coefficients B and D in the transformed Eq. (7.165) for further derivations of the matrix adjoint problem. Also, we need to interchange the lower two rows of the matrix coefficient C in the matrix IVC, Eq. (7.167), and multiply the resulting lower row by $-\alpha$ to make this coefficient identical to the matrix coefficient C in the transformed Eq. (7.165).

The resulting matrix BCs and IVC obtain the form:

$$\begin{pmatrix} 0 & 0 & 0 \\ 0 & 1 & 0 \\ 0 & 0 & 0 \end{pmatrix} \begin{pmatrix} v \\ u \\ \Gamma \end{pmatrix}\bigg|_a = \begin{pmatrix} 0 \\ f_a \\ 0 \end{pmatrix}, \quad \begin{pmatrix} 0 & 0 & 0 \\ 0 & 1 & 0 \\ 0 & 0 & 0 \end{pmatrix} \begin{pmatrix} v \\ u \\ \Gamma \end{pmatrix}\bigg|_b = \begin{pmatrix} 0 \\ f_b \\ 0 \end{pmatrix} \quad (7.168)$$

$$\begin{pmatrix} 0 & 1 & 0 \\ 0 & 0 & 0 \\ -\alpha & 0 & 0 \end{pmatrix} \begin{pmatrix} v \\ u \\ \Gamma \end{pmatrix}\bigg|_{t=0} = \begin{pmatrix} f_0 \\ 0 \\ -\alpha f_1 \end{pmatrix} \quad (7.169)$$

The corresponding matrix adjoint problem has the form of Eqs. (7.26)–(7.28). To formulate its right-hand term, we assume the observable R in the form of Eq. (7.43). Then, from comparison with the general expression Eq. (7.44) we conclude that the matrix observables weighting function W has here the form of a column vector

$$W = \begin{pmatrix} 0 \\ p \\ 0 \end{pmatrix} \quad (7.170)$$

Making a realistic assumption that the wave propagation speed does not depend on time, and repeating the pattern of derivations of the previous Subsection, we obtain the sought for matrix adjoint problem in the expanded form:

$$-\begin{pmatrix} 0 & 0 & -\alpha \\ 1 & 0 & 0 \\ 0 & 0 & 0 \end{pmatrix}\begin{pmatrix} \dot{v}^* \\ \dot{u}^* \\ \dot{\Gamma}^* \end{pmatrix} - \begin{pmatrix} 0 & 0 & 0 \\ 0 & 1 & 0 \\ 0 & 0 & 1 \end{pmatrix}\begin{pmatrix} v^{*\prime} \\ u^{*\prime} \\ \Gamma^{*\prime} \end{pmatrix} + \begin{pmatrix} -1 & 0 & 0 \\ 0 & 0 & 0 \\ 0 & -1 & 0 \end{pmatrix}\begin{pmatrix} v^* \\ u^* \\ \Gamma^* \end{pmatrix} = \begin{pmatrix} 0 \\ p \\ 0 \end{pmatrix}$$

$$(7.171)$$

$$\begin{pmatrix} 0 & 0 & 0 \\ 0 & 0 & 0 \\ 0 & 0 & -1 \end{pmatrix}\begin{pmatrix} v^* \\ u^* \\ \Gamma^* \end{pmatrix}\Bigg|_a = \begin{pmatrix} 0 \\ 0 \\ 0 \end{pmatrix}, \quad \begin{pmatrix} 0 & 0 & 0 \\ 0 & 0 & 0 \\ 0 & 0 & -1 \end{pmatrix}\begin{pmatrix} v^* \\ u^* \\ \Gamma^* \end{pmatrix}\Bigg|_b = \begin{pmatrix} 0 \\ 0 \\ 0 \end{pmatrix} \quad (7.172)$$

$$\begin{pmatrix} 0 & 0 & -\alpha \\ 1 & 0 & 0 \\ 0 & 0 & 0 \end{pmatrix}\begin{pmatrix} v^* \\ u^* \\ \Gamma^* \end{pmatrix}\Bigg|_{t=T} = \begin{pmatrix} 0 \\ 0 \\ 0 \end{pmatrix} \quad (7.173)$$

Rewriting the Eqs. (7.171)–(7.173) in the form of a system of scalar PDEs with corresponding BCs and a FVCs and discarding the indefinite BCs and FVC, we have:

$$\alpha\dot{\Gamma}^* - v^* = 0 \quad (7.174)$$

$$-\dot{v}^* - u^{*\prime} = p \quad (7.175)$$

$$-\Gamma^{*\prime} - u^* = 0 \quad (7.176)$$

$$\Gamma^*|_a = 0, \quad \Gamma^*|_b = 0 \quad (7.177)$$

$$\Gamma^*|_{t=T} = 0, \quad \dot{\Gamma}^*\Big|_{t=T} = 0 \quad (7.178)$$

[In the second FVC of Eq. (7.178) we made use of Eq. (7.174)]. From Eqs. (7.174) and (7.176) we have:

$$-\dot{v}^* = -\alpha\ddot{\Gamma}^*, \quad -u^{*\prime} = \Gamma^{*\prime\prime} \quad (7.179)$$

Here we made a realistic assumption that the propagation speed of the wave does not depend on time. Substituting in Eq. (7.175), we obtain the scalar adjoint wave equation:

$$-\alpha\ddot{\Gamma}^* + \Gamma^{*\prime\prime} = p \quad (7.180)$$

Introducing the scalar adjoint solution $w = \Gamma^*$, we can rewrite the PDE Eq. (7.180), BCs Eq. (7.177) and FVCs Eq. (7.178) in the form of the scalar adjoint problem corresponding to the forward problem, Eqs. (7.159)–(7.161):

$$-\frac{1}{c^2}\ddot{w} + w'' = p \qquad (7.181)$$

$$w|_a = 0, \quad w|_b = 0 \qquad (7.182)$$

$$w|_{t=T} = 0, \quad \dot{w}\Big|_{t=T} = 0 \qquad (7.183)$$

where in the scalar PDE, Eq. (7.181), we returned to the wave propagation speed via the relation $\alpha = 1/c^2$.

All forward problems considered in this chapter so far were limited to 1D space. Now we move on to higher-dimensional spaces.

7.4 Stationary and Non-stationary Problems in 2D and 3D Space

7.4.1 Poisson Equation

We consider a forward problem with the Poisson equation and the Dirichlet BC, Eqs. (7.31) and (7.32), where the Laplacian Δ in the equation is represented as a product of two ∇—operators:

$$\nabla^T \nabla u = f(\mathbf{r}) \qquad (7.184)$$

$$u|_S = f_S(\mathbf{r}_S) \qquad (7.185)$$

To reduce the PDE of 2nd order Eq. (7.184) to a system of PDEs of 1st order we introduce the additional dependent variable, the vector $\mathbf{\Gamma} = \nabla u$. Then, Eq. (7.184) becomes a system of two PDEs of 1st order:

$$\nabla u - \mathbf{\Gamma} = \mathbf{0} \qquad (7.186)$$

$$\nabla^T \mathbf{\Gamma} = f \qquad (7.187)$$

The system Eqs. (7.186) and (7.187) is further combined into a single matrix ODE of 1st order similar to Eq. (7.2). In expanded form we have:

$$\begin{pmatrix} \nabla & \mathbf{O} \\ 0 & \nabla^T \end{pmatrix} \begin{pmatrix} u \\ \mathbf{\Gamma} \end{pmatrix} + \begin{pmatrix} 0 & -\mathbf{I} \\ 0 & \mathbf{0}^T \end{pmatrix} \begin{pmatrix} u \\ \mathbf{\Gamma} \end{pmatrix} = \begin{pmatrix} \mathbf{0} \\ f \end{pmatrix} \qquad (7.188)$$

Note that in Eq. (7.188), the analog of the matrix coefficient D in Eq. (7.2) becomes a matrix differential operator.

The form of Eq. (7.188) implies that the BC imposed on the component u has to have a vector form. The natural way to implement this requirement is to multiply both sides of Eq. (7.185) by a unit vector $\mathbf{n}$, normal to the boundary S at a given point $\mathbf{r}_S$:

$$\mathbf{n}u|_S = \mathbf{n}f_S \tag{7.189}$$

To ensure that the component Γ does not constrain the forward solution u, we impose on it an indefinite BC in the form:

$$\mathbf{0}^T\Gamma|_S = 0 \tag{7.190}$$

The resulting matrix forward problem in the expanded form is as follows:

$$\begin{cases} \begin{pmatrix} \nabla & \mathbf{O} \\ 0 & \nabla^T \end{pmatrix} \begin{pmatrix} u \\ \Gamma \end{pmatrix} + \begin{pmatrix} \mathbf{0} & -\mathbf{I} \\ 0 & \mathbf{0}^T \end{pmatrix} \begin{pmatrix} u \\ \Gamma \end{pmatrix} = \begin{pmatrix} \mathbf{0} \\ f \end{pmatrix} \\ \begin{pmatrix} \mathbf{n} & \mathbf{O} \\ 0 & \mathbf{0}^T \end{pmatrix} \begin{pmatrix} u \\ \Gamma \end{pmatrix}_S = \begin{pmatrix} \mathbf{n}f_S \\ 0 \end{pmatrix} \end{cases} \tag{7.191}$$

Thus, it has the form similar to that of the system Eqs. (7.2) and (7.3):

$$\begin{cases} DX + AX = S_e \\ BX|_S = S_S \end{cases} \tag{7.192}$$

where

$$D = \begin{pmatrix} \nabla & \mathbf{O} \\ 0 & \nabla^T \end{pmatrix}; \, A = \begin{pmatrix} \mathbf{0} & -\mathbf{I} \\ 0 & \mathbf{0}^T \end{pmatrix}; \, S_e = \begin{pmatrix} \mathbf{0} \\ f \end{pmatrix}; \, B = \begin{pmatrix} \mathbf{n} & \mathbf{O} \\ 0 & \mathbf{0}^T \end{pmatrix}; \, S_S = \begin{pmatrix} \mathbf{n}f_S \\ 0 \end{pmatrix} \tag{7.193}$$

And, in matrix operator form we have the higher-dimension analogue of Eq. (7.4):

$$DX + AX - \delta(\mathbf{r} - \mathbf{r}_S)BX = S_e - \delta(\mathbf{r} - \mathbf{r}_S)S_S \tag{7.194}$$

Now we derive the matrix adjoint operator using the Lagrange identity, where the inner products $(X*, LX)$ and (L^*X^*, X) are understood as integrals over the 2D or 3D domain V. For the 2nd and 3rd terms in the left side of Eq. (7.194), we immediately obtain:

$$\int_V X^{*T} AX \, dV = \int_V (A^T X^*)^T X dV \tag{7.195}$$

$$\int_V X^{*T} \delta(\mathbf{r} - \mathbf{r}_S) BX \ dV = \int_S X^{*T} BX \ dS = \int_S (B^T X^*)^T X \ dS \qquad (7.196)$$

For the first term in the left side of Eq. (7.194), we consider the 2D and 3D cases separately.

In the 2D case, representing the integration over the domain V explicitly by integration over x and y, we have:

$$\int_V X^{*T} DX \ dV = \underline{\int dx \int dy}_V \begin{pmatrix} \mathbf{u}^* \\ \mathbf{\Gamma}^* \end{pmatrix}^T \begin{pmatrix} \nabla & \mathbf{O} \\ 0 & \nabla^T \end{pmatrix} \begin{pmatrix} u \\ \mathbf{\Gamma} \end{pmatrix}$$

$$= \underline{\int dx \int dy}_V \begin{pmatrix} \mathbf{u}^* \\ \mathbf{\Gamma}^* \end{pmatrix}^T \begin{pmatrix} \nabla u \\ \nabla^T \mathbf{\Gamma} \end{pmatrix} = \underline{\int dx \int dy}_V (\mathbf{u}^{*T} \nabla u + \mathbf{\Gamma}^* \nabla^T \mathbf{\Gamma})$$

$$(7.197)$$

Representing all vector quantities through their x- and y-components, denoting $\partial_x = \partial/\partial x$, $\partial_y = \partial/\partial y$, and re-grouping the terms, we have:

$$\int_V X^{*T} DX \ dV = \underline{\int dy \int dx}_V \left(u_x^* \partial_x u + \mathbf{\Gamma}^* \partial_x \mathbf{\Gamma}_x \right) + \underline{\int dx \int dy}_V \left(u_y^* \partial_y u + \mathbf{\Gamma}^* \partial_y \mathbf{\Gamma}_y \right)$$

$$(7.198)$$

Integrating by parts the inner integrals in the right side of Eq. (7.198), we have (cf. Fig. 7.1):

$$\underline{\int dy \int dx}_V \left(u_x^* \partial_x u + \mathbf{\Gamma}^* \partial_x \mathbf{\Gamma}_x \right) = \int_{y_{min}}^{y_{max}} dy \left(u_x^* u + \mathbf{\Gamma}^* \mathbf{\Gamma}_x \right) \Big|_{x_1(y)}^{|x_2(y)}$$

$$- \int_V dV \left(\partial_x u_x^* u + \partial_x \mathbf{\Gamma}^* \mathbf{\Gamma}_x \right) \qquad (7.199)$$

$$\underline{\int dx \int dy}_V \left(u_y^* \partial_y u + \mathbf{\Gamma}^* \partial_y \mathbf{\Gamma}_y \right) = \int_{x_{min}}^{x_{max}} dx \left(u_y^* u + \mathbf{\Gamma}^* \mathbf{\Gamma}_y \right) \Big|_{y_1(x)}^{|y_2(x)}$$

$$- \int_V dV \left(\partial_y u_y^* u + \partial_y \mathbf{\Gamma}^* \mathbf{\Gamma}_y \right) \qquad (7.200)$$

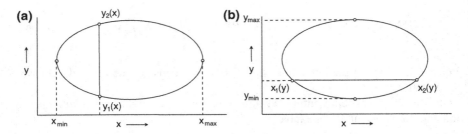

Fig. 7.1 Limits of integration over the 2D domain: **a** in Eq. (7.199), **b** in Eq. (7.200)

Substituting Eqs. (7.199) and (7.200) in Eq. (7.198) and re-grouping the terms, we have:

$$\int_V X^{*T} DX \, dV = - \int_V dV \left[\left(\partial_x u_x^* + \partial_y u_y^* \right) u + \partial_x \Gamma^* \Gamma_x + \partial_y \Gamma^* \Gamma_y \right]$$

$$+ \int_{y_{min}}^{y_{max}} dy \left(u_x^* u + \Gamma^* \Gamma_x \right) \Big|_{x_1(y)}^{x_2(y)} + \int_{x_{min}}^{x_{max}} dx \left(u_y^* u + \Gamma^* \Gamma_y \right) \Big|_{y_1(x)}^{y_2(x)}$$

$$(7.201)$$

Now, representing the outer normal **n** to the boundary contour S in the form (cf. Fig. 7.2):

$$\mathbf{n} = \begin{pmatrix} n_x \\ n_y \end{pmatrix} = \begin{pmatrix} dy/dS \\ -dx/dS \end{pmatrix}, \qquad (7.202)$$

and substituting $dx = -n_y dS$ and $dy = n_x dS$ into integrals over x and y, we can re-write Eq. (7.201) in a vector form:

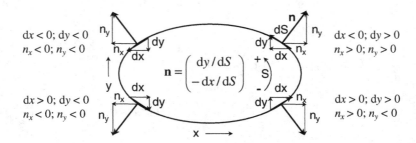

Fig. 7.2 Definition of the outward normal **n** to the boundary S of a 2D domain using components dx and dy of the differential dS. By absolute value, $|n_x| = |dy/dS|$, and $|n_y| = |dx/dS|$. With the chosen counterclockwise direction of integration along the boundary, $n_x = dy/dS$, and $n_y = -dx/dS$

$$\int_V X^{*T} DX \, dV = -\int_V dV\left[(\nabla^T \mathbf{u}^*)\, u + (\nabla \Gamma^*)^T \Gamma\right]$$

$$+ \int_S dS\left(\mathbf{n}^T \mathbf{u}^* u + \Gamma^* \mathbf{n}^T \Gamma\right) \tag{7.203}$$

$$= -\int_V \begin{pmatrix} \nabla^T \mathbf{u}^* \\ \nabla \Gamma^* \end{pmatrix}^T \begin{pmatrix} u \\ \Gamma \end{pmatrix} dV + \int_S \begin{pmatrix} \mathbf{n}^T \mathbf{u}^* \\ \mathbf{n}\Gamma^* \end{pmatrix}^T \begin{pmatrix} u \\ \Gamma \end{pmatrix} dS$$

In the 3D case, representing the integration over the domain V explicitly, by integration over x, y, and z we have:

$$\int_V X^{*T} DX \, dV = \underline{\int dx \int dy \int dz}_{V} \begin{pmatrix} \mathbf{u}^* \\ \Gamma^* \end{pmatrix}^T \begin{pmatrix} \nabla u \\ \nabla^T \Gamma \end{pmatrix}$$

$$= \underline{\int dx \int dy \int dz}_{V} (\mathbf{u}^{*T}\nabla u + \Gamma^* \nabla^T \Gamma)$$

$$= \underline{\int dz \int dy \int dx}_{V} \left(u_x^* \partial_x u + \Gamma^* \partial_x \Gamma_x\right) \tag{7.204}$$

$$+ \underline{\int dz \int dx \int dy}_{V} \left(u_y^* \partial_y u + \Gamma^* \partial_y \Gamma_y\right)$$

$$+ \underline{\int dz \int dy \int dx}_{V} \left(u_z^* \partial_z u + \Gamma^* \partial_z \Gamma_z\right)$$

Representing all vector quantities through their x-, y- and z-components, denoting $\partial_z = \partial/\partial z$, re-grouping the terms, and integrating by parts the inner integrals in the right side of Eq. (7.204), we have:

$$\underline{\int dy \int dz \int dx}_{V} \left(u_x^* \partial_x u + \Gamma^* \partial_x \Gamma_x\right) = \underline{\int dy \int dz}_{V_{YZ}} \left.\left(u_x^* u + \Gamma^* \Gamma_x\right)\right|_{x_1(y,z)}^{x_2(y,z)}$$

$$- \int_V dV\left(\partial_x u_x^* u + \partial_x \Gamma^* \Gamma_x\right) \tag{7.205}$$

$$\underline{\int dx \int dz \int dy}_{V} \left(u_y^* \partial_y u + \Gamma^* \partial_y \Gamma_y\right) = \underline{\int dx \int dz}_{V_{XZ}} \left.\left(u_y^* u + \Gamma^* \Gamma_y\right)\right|_{y_1(x,z)}^{y_2(x,z)}$$

$$- \int_V dV\left(\partial_y u_y^* u + \partial_y \Gamma^* \Gamma_y\right) \tag{7.206}$$

$$\int_V dx \int dy \int dz \left(u_z^* \partial_z u + \Gamma^* \partial_z \Gamma_z\right) = \int dx \int dy \left(u_z^* u + \Gamma^* \Gamma_z\right)\Big|_{z_1(x,y)}^{z_2(x,y)}$$

$$- \int_V dV \left(\partial_z u_z^* u + \partial_z \Gamma^* \Gamma_z\right) \tag{7.207}$$

where V_{YZ}, V_{XZ}, and V_{XY} are 2D projections of the volume V at planes of coordinates (y,z), (x,z), and (x,y) respectively. Substituting Eqs. (7.205)–(7.207) in Eq. (7.204) and re-grouping the terms, we have:

$$\int_V X^{*T} DX\, dV = -\int_V dV \left[\left(\partial_x u_x^* + \partial_y u_y^* + \partial_z u_z^*\right) u + \partial_x \Gamma^* \Gamma_x + \partial_y \Gamma^* \Gamma_y + \partial_z \Gamma^* \Gamma_z\right]$$

$$+ \int dy \int dz \left(u_x^* u + \Gamma^* \Gamma_x\right)\Big|_{x_1(y,z)}^{x_2(y,z)}$$
$$\underbrace{}_{V_{YZ}}$$

$$+ \underbrace{\int dx \int dz \left(u_y^* u + \Gamma^* \Gamma_y\right)\Big|_{y_1(x,z)}^{y_2(x,z)}}_{V_{XZ}} + \underbrace{\int dx \int dy \left(u_z^* u + \Gamma^* \Gamma_z\right)\Big|_{z_1(x,y)}^{z_2(x,y)}}_{V_{XY}}$$

$$\tag{7.208}$$

Now we represent the outer normal $\mathbf{n}$ to the boundary surface S in the form

$$\mathbf{n} = \begin{pmatrix} n_x \\ n_y \\ n_z \end{pmatrix} = \begin{pmatrix} \pm dy \cdot dz/dS \\ \pm dx \cdot dz/dS \\ \pm dx \cdot dy/dS \end{pmatrix}, \tag{7.209}$$

where dS is the element of the boundary dx, dy, and dz are its projections at the corresponding axes of coordinates, and the signs of the components at a given point of the boundary are chosen to keep the normal $\mathbf{n}$ pointing outward. Substituting $dydz = \pm n_x dS$, $dxdz = \pm n_y dS$ and $dxdy = \pm n_z dS$, we can re-write Eq. (7.208) in the same vector form as Eq. (7.203). Further on, representing

$$\begin{pmatrix} \nabla^T \mathbf{u}^* \\ \nabla \Gamma^* \end{pmatrix} = \begin{pmatrix} \nabla^T & \mathbf{0} \\ 0 & \nabla \end{pmatrix}\begin{pmatrix} \mathbf{u}^* \\ \Gamma^* \end{pmatrix}, \quad \text{and} \quad \begin{pmatrix} \mathbf{n}^T \mathbf{u}^* \\ \mathbf{n}\Gamma^* \end{pmatrix} = \begin{pmatrix} \mathbf{n}^T & \mathbf{0} \\ 0 & \mathbf{n} \end{pmatrix}\begin{pmatrix} \mathbf{u}^* \\ \Gamma^* \end{pmatrix}$$

for both 2D and 3D cases, we obtain

$$\int_V X^{*T} DX\, dV = -\int_V \left[\begin{pmatrix} \nabla^T & \mathbf{0} \\ 0 & \nabla \end{pmatrix}\begin{pmatrix} \mathbf{u}^* \\ \Gamma^* \end{pmatrix}\right]^T \begin{pmatrix} u \\ \Gamma \end{pmatrix} dV + \int_S \left[\begin{pmatrix} \mathbf{n}^T & \mathbf{0} \\ 0 & \mathbf{n} \end{pmatrix}\begin{pmatrix} \mathbf{u}^* \\ \Gamma^* \end{pmatrix}\right]^T \begin{pmatrix} u \\ \Gamma \end{pmatrix} dS$$

$$= -\int_V (D^T X^*)^T X dV + \int_S (JX^*)^T X dS$$

$$\tag{7.210}$$

where

$$J = \begin{pmatrix} \mathbf{n}^T & 0 \\ \mathbf{O} & \mathbf{n} \end{pmatrix} \tag{7.211}$$

is a quasi-identity matrix, a higher-dimensional analog of the identity matrix. Summing Eqs. (7.210), (7.195) and (7.196) together, we obtain:

$$(X^*, LX) = \int_V \left(-D^T X^* + A^T X^* \right)^T X \, dV - \int_S \left[(B^T - J) X^* \right]^T X \, dS \tag{7.212}$$

Resulting adjoint matrix operator has the form:

$$L^* X^* = -D^T X^* + A^T X^* - \delta(\mathbf{r} - \mathbf{r}_S) \left(B^T - J \right) X^* \tag{7.213}$$

And we obtain the corresponding matrix adjoint problem in the form

$$\begin{cases} -D^T X^* + A^T X^* = W_e \\ (B^T - J) X^* |_S = W_c \end{cases} \tag{7.214}$$

We assume the observable R in the form analogous to Eqs. (7.43) and (7.44):

$$R = \int_V p(\mathbf{r}) u(\mathbf{r}) \, dV = (p, u) \tag{7.215}$$

$$R = \int_V W^T(\mathbf{r}) X(\mathbf{r}) \, dV = (W, X) \tag{7.216}$$

where the matrix observables weighting function W has the form of a column block vector

$$W = \begin{pmatrix} p \\ \mathbf{0} \end{pmatrix} \tag{7.217}$$

In the expanded form we have:

$$\begin{cases} -\begin{pmatrix} \nabla^T & 0 \\ \mathbf{O} & \nabla \end{pmatrix} \begin{pmatrix} \mathbf{u}^* \\ \Gamma^* \end{pmatrix} + \begin{pmatrix} 0^T & 0 \\ -\mathbf{I} & 0 \end{pmatrix} \begin{pmatrix} \mathbf{u}^* \\ \Gamma^* \end{pmatrix} = \begin{pmatrix} p \\ \mathbf{0} \end{pmatrix} \\ \begin{pmatrix} 0^T & 0 \\ \mathbf{O} & -\mathbf{n} \end{pmatrix} \begin{pmatrix} \mathbf{u}^* \\ \Gamma^* \end{pmatrix}_S = \begin{pmatrix} 0 \\ \mathbf{0} \end{pmatrix} \end{cases} \tag{7.218}$$

The corresponding system of 1st order PDEs with BCs has the form:

$$\begin{cases} -\nabla^T \mathbf{u}^* = p \\ -\nabla \Gamma^* - \mathbf{u}^* = \mathbf{0} \\ \left(\mathbf{0}^T \mathbf{u}^* + 0 \cdot \Gamma^* \right)\big|_S = 0 \\ \mathbf{n}\Gamma^*\big|_S = \mathbf{0} \end{cases} \tag{7.219}$$

Applying the operator ∇^T to the 2nd equation of the system Eq. (7.219), and substituting the result in the 1st equation of this system, we obtain a PDE with respect to a scalar function Γ^*:

$$\nabla^T \nabla \Gamma^* = p \tag{7.220}$$

Discarding the indefinite BC in the system Eq. (7.219), we rewrite the remaining BC in this system in the form:

$$\Gamma^*\big|_S = 0 \tag{7.221}$$

And, introducing the scalar adjoint solution $w = \Gamma^*$, we obtain the scalar adjoint problem corresponding to the scalar forward problem, Eqs. (7.184), (7.185):

$$\nabla^T \nabla w = p \tag{7.222}$$

$$w\big|_S = 0 \tag{7.223}$$

7.4.2 Wave Equation

We consider a forward problem with the wave equation, Dirichlet BC, and an IVC, Eqs. (7.156)–(7.158). Representing the Laplacian Δ in the equation as a product of two ∇—operators, and denoting again $\alpha \equiv 1/c^2$, we have:

$$-\alpha(\mathbf{r}, t)\, \ddot{u} + \nabla^T \nabla u = f(\mathbf{r}, t) \tag{7.224}$$

$$u\big|_S = f_S(\mathbf{r}_S, t) \tag{7.225}$$

$$u\big|_{t=0} = f_0(\mathbf{r}), \quad \dot{u}\big|_{t=0} = f_1(\mathbf{r}) \tag{7.226}$$

To reduce the PDE of 2nd order Eq. (7.224) to a system of PDEs of 1st order we introduce additional dependent variables $\upsilon = \dot{u}$ and $\mathbf{\Gamma} = \nabla u$. Then, Eq. (7.224) becomes a system of three PDEs:

$$\dot{u} - v = 0 \tag{7.227}$$

$$\nabla u - \mathbf{\Gamma} = \mathbf{0} \tag{7.228}$$

$$-\alpha \dot{v} + \nabla^T \mathbf{\Gamma} = f \tag{7.229}$$

The BC imposed on u and an indefinite BC imposed on $\mathbf{\Gamma}$ in the form of Eqs. (7.189) and (7.190) are complemented by an indefinite BC for the additional dependent variable v:

$$0 \cdot v|_S = 0 \tag{7.230}$$

The IVCs, Eq. (7.226), imposed on u and v are complemented by an indefinite IVC for the additional dependent variable $\mathbf{\Gamma}$:

$$\mathbf{0}\mathbf{\Gamma}|_{t=0} = \mathbf{0} \tag{7.231}$$

To proceed, we need to perform transformations of the resulting matrix BC and IVC as we did in the 1D case with Eqs. (7.166) and (7.167). The resulting matrix forward problem in the expanded form is as follows:

$$\begin{cases} \begin{pmatrix} 0 & 1 & \mathbf{0}^T \\ \mathbf{0} & 0 & \mathbf{O} \\ -\alpha & 0 & \mathbf{0}^T \end{pmatrix} \begin{pmatrix} \dot{v} \\ \dot{u} \\ \dot{\mathbf{\Gamma}} \end{pmatrix} + \begin{pmatrix} 0 & 0 & \mathbf{0}^T \\ \mathbf{0} & \nabla & \mathbf{O} \\ 0 & 0 & \nabla^T \end{pmatrix} \begin{pmatrix} v \\ u \\ \mathbf{\Gamma} \end{pmatrix} + \begin{pmatrix} -1 & 0 & \mathbf{0}^T \\ \mathbf{0} & 0 & -\mathbf{I} \\ 0 & 0 & \mathbf{0}^T \end{pmatrix} \begin{pmatrix} v \\ u \\ \mathbf{\Gamma} \end{pmatrix} = \begin{pmatrix} 0 \\ \mathbf{0} \\ f \end{pmatrix} \\ \\ \begin{cases} \begin{pmatrix} 0 & 0 & \mathbf{0}^T \\ \mathbf{0} & \mathbf{n} & \mathbf{O} \\ 0 & 0 & \mathbf{0}^T \end{pmatrix} \begin{pmatrix} v \\ u \\ \mathbf{\Gamma} \end{pmatrix}_S = \begin{pmatrix} 0 \\ \mathbf{n} f_S \\ 0 \end{pmatrix}; \begin{pmatrix} 0 & 1 & \mathbf{0}^T \\ \mathbf{0} & 0 & \mathbf{O} \\ -\alpha & 0 & \mathbf{0}^T \end{pmatrix} \begin{pmatrix} v \\ u \\ \mathbf{\Gamma} \end{pmatrix}_{t=0} = \begin{pmatrix} f_0 \\ \mathbf{0} \\ -\alpha f_1 \end{pmatrix} \end{cases} \end{cases} \tag{7.232}$$

Thus, it has the form similar to that of the system of Eqs. (7.17)–(7.19):

$$\begin{cases} C\dot{X} + DX + AX = S_e \\ BX|_S = S_S; CX|_{t=0} = S_0 \end{cases} \tag{7.233}$$

where the block matrix C in the matrix ODE and IVC has the form:

$$C = \begin{pmatrix} 0 & 1 & \mathbf{0}^T \\ \mathbf{0} & 0 & \mathbf{O} \\ -\alpha & 0 & \mathbf{0}^T \end{pmatrix} \tag{7.234}$$

And, in matrix operator form we obtain the higher-dimensional analogue of Eq. (7.20):

$$C\dot{X} + DX + AX - \delta(\mathbf{r} - \mathbf{r}_S)BX + \delta(t)CX = S_e - \delta(\mathbf{r} - \mathbf{r}_S)S_S + \delta(t)S_0 \quad (7.235)$$

Now we derive the matrix adjoint operator using the Lagrange identity, where the inner products (X^*, LX) and (L^*X^*, X) are understood as integrals over the 2D or 3D domain V complemented by integration over the time interval of integration $[0, T]$. Once again, we make a realistic assumption that the wave propagation speed does not depend on time. For the 3rd and 4th terms in the left side of Eq. (7.235), importing the results of the previous Subsection, we have [cf. Eq. (7.195) and (7.196)]:

$$\int_0^T dt \int_V dV\, X^{*T}AX = \int_0^T dt \int_V dV (A^T X^*)^T X \quad (7.236)$$

$$\int_V X^{*T}\delta(\mathbf{r} - \mathbf{r}_S)BX\, dS = \int_0^T dt \int_S dS\, X^{*T}BX = \int_0^T dt \int_S dS (B^T X^*)^T X \quad (7.237)$$

For the 2nd term, following the pattern of derivations in the previous section, we have [cf. Eq. (7.210)]:

$$\int_0^T dt \int_V dV\, X^{*T}DX = -\int_0^T dt \int_V dV\, (D^T X^*)^T X + \int_0^T dt \int_S dS\, (JX^*)^T X \quad (7.238)$$

where [compare with Eq. (7.211)]

$$J = \begin{pmatrix} 0 & \mathbf{0}^T & 0 \\ 0 & \mathbf{n}^T & 0 \\ \mathbf{0} & \mathbf{O} & \mathbf{n} \end{pmatrix} \quad (7.239)$$

Finally, the 1st term is handled by integration by parts over t:

$$\int_0^T dt \int_V dV\, X^{*T}C\dot{X} = -\int_0^T dt \int_V dV\, \left(C^T \dot{X^*}\right)^T X + \int_V dV\, [C^T X^*]^T X \Big|_{t=0}^{t=T} \quad (7.240)$$

Substituting Eqs. (7.236)–(7.238) and (7.240) into Eq. (7.235), we obtain:

$$(X^*, LX) = \int_0^T dt \int_V dV \left(-C^T \dot{X^*} - D^T X^* + A^T X^*\right)^T X$$

$$- \int_0^T dt \int_S dS\, [(B^T - J)X^*]^T X + \int_V dV (C^T X^*)^T X \Big|_{t=T} \quad (7.241)$$

The resulting adjoint matrix operator equation has the form:

$$L^* X^* = -C^T \dot{X}^* - D^T X^* + A^T X^* - \delta(\mathbf{r} - \mathbf{r}_S)(B^T - J)X^* + \delta(t - T)C^T X^*$$

$$(7.242)$$

And we obtain the resulting matrix adjoint problem in the form

$$\begin{cases} -C^T \dot{X}^* - D^T X^* + A^T X^* = W_e \\ \quad (B^T - J)X^*|_S = W_S \\ \quad\quad C^T X^*|_{t=T} = W_T \end{cases} \qquad (7.243)$$

Assuming the observable R as defined in the form analogous to Eqs. (7.215) and (7.216)

$$R = \int_0^T dt \int_V dV p(\mathbf{r}, t) u(\mathbf{r}, t) = (p, u) \qquad (7.244)$$

$$R = \int_0^T dt \int_V dV \, W^T(\mathbf{r}, t) X(\mathbf{r}, t) = (W, X) \qquad (7.245)$$

in the expanded form we have:

$$\begin{cases} -\begin{pmatrix} 0 & \mathbf{0}^T & -\alpha \\ 1 & \mathbf{0}^T & 0 \\ \mathbf{0} & \mathbf{O} & \mathbf{0} \end{pmatrix}\begin{pmatrix} \dot{v}^* \\ \dot{\mathbf{u}}^* \\ \dot{\boldsymbol{\Gamma}}^* \end{pmatrix} - \begin{pmatrix} 0 & \mathbf{0}^T & 0 \\ 0 & \nabla^T & 0 \\ \mathbf{0} & \mathbf{O} & \nabla \end{pmatrix}\begin{pmatrix} v^* \\ \mathbf{u}^* \\ \boldsymbol{\Gamma}^* \end{pmatrix} + \begin{pmatrix} -1 & \mathbf{0}^T & 0 \\ 0 & \mathbf{0}^T & 0 \\ \mathbf{0} & -\mathbf{I} & \mathbf{0} \end{pmatrix}\begin{pmatrix} v^* \\ \mathbf{u}^* \\ \boldsymbol{\Gamma}^* \end{pmatrix} = \begin{pmatrix} 0 \\ p \\ \mathbf{0} \end{pmatrix} \\ \begin{pmatrix} 0 & \mathbf{0}^T & 0 \\ 0 & \mathbf{0}^T & 0 \\ \mathbf{0} & \mathbf{O} & -\mathbf{n} \end{pmatrix}\begin{pmatrix} v^* \\ \mathbf{u}^* \\ \boldsymbol{\Gamma}^* \end{pmatrix}_S = \begin{pmatrix} 0 \\ 0 \\ \mathbf{0} \end{pmatrix} ; \begin{pmatrix} 0 & \mathbf{0}^T & -\alpha \\ 1 & \mathbf{0}^T & 0 \\ \mathbf{0} & \mathbf{O} & \mathbf{0} \end{pmatrix}\begin{pmatrix} v^* \\ \mathbf{u}^* \\ \boldsymbol{\Gamma}^* \end{pmatrix}_{t=T} = \begin{pmatrix} 0 \\ 0 \\ \mathbf{0} \end{pmatrix} \end{cases}$$

$$(7.246)$$

Rewriting the system Eq. (7.246) in the form of a system of scalar PDEs with corresponding BCs and an FVC and discarding the indefinite BCs and FVC, we have:

$$\alpha \, \dot{\boldsymbol{\Gamma}}^* - v^* = 0 \qquad (7.247)$$

$$-\dot{v}^* - \nabla^T \mathbf{u}^* = p \qquad (7.248)$$

$$-\nabla \boldsymbol{\Gamma}^* - \mathbf{u}^* = \mathbf{0} \qquad (7.249)$$

$$\mathbf{n}\Gamma^*|_a = 0, \quad \mathbf{n}\Gamma^*|_b = 0 \tag{7.250}$$

$$\Gamma^*|_{t=T} = 0, \quad \dot{\Gamma}^*\bigg|_{t=T} = 0 \tag{7.251}$$

[In the second FVC of Eq. (7.251) we made use of Eq. (7.247)]. From Eqs. (7.247) and (7.249) we have:

$$\dot{v}^* - \alpha \ddot{\Gamma}^*, \quad -\nabla^T \mathbf{u}^* = \nabla^T \nabla \Gamma^* \tag{7.252}$$

Substituting in Eq. (7.248), we obtain the scalar adjoint wave equation:

$$-\alpha \ddot{\Gamma}^* + \nabla^T \nabla \Gamma^* = p \tag{7.253}$$

Introducing the scalar adjoint solution $w = \Gamma^*$, we can rewrite the PDE Eq. (7.253), BCs Eq. (7.250) and FVC Eq. (7.251) in the form of the adjoint problem corresponding to the forward problem, Eqs. (7.224)–(7.226):

$$-\frac{1}{c^2} \ddot{w} + \nabla^T \nabla w = p \tag{7.254}$$

$$w|_S = 0 \tag{7.255}$$

$$w|_{t=T} = 0, \quad \dot{w}\bigg|_{t=T} = 0 \tag{7.256}$$

where in the PDE, Eq. (7.254), we returned to wave propagation speed via the relation $\alpha = 1/c^2$.

References

The material presented in this chapter was partly developed by the author during preparation of the course on sensitivity analysis presented at the Jet Propulsion Laboratory some time ago. The rest of the material is new and is published in this book for the first time. The technique of reducing the higher-order differential equations to systems of first-order differential equations used in this chapter is well-known in the literature and can be found in any suitable book on differential equations. The conventional approach to formulation of adjoint problems with higher-order differential equations is presented in detail in the monograph written by Gury Marchuk (1995)

Marchuk GI (1995) Adjoint equations and analysis of complex systems. Kluver Academic Publishers, Dordrecht, Boston, London

Chapter 8
Applications of Sensitivity Analysis in Remote Sensing

Abstract In this chapter we consider the practical applications of sensitivity analysis in remote sensing. After a brief review of various types of sensitivities, we consider three main areas of applications: the error analyses of input and output parameters and the solution of inverse problems. The error analysis of output parameters with given errors of input parameters is most straightforward. The corresponding algorithm involves, essentially, only matrix multiplication. The error analysis of input parameters with given requirements to errors of output parameters becomes more complicated if the matrix of sensitivities cannot be inverted directly. The solution of inverse problems is the most sophisticated area of application of sensitivity analysis. Here, the simple forms of least squares method and of the method of statistical regularization are presented.

Keywords Random variables · Error analysis · Inverse problems

8.1 Sensitivities of Models: A Summary

As stated in Chap. 2, in general, both input and output parameters of models used in remote sensing can be of two types: discrete parameters and continuous parameters. The latter, by their nature are functions of some arguments, which are considered to be independent variables. In this Section we consider both types of parameters, as well as sensitivities to discrete and continuous parameters.

8.1.1 Discrete Parameters and Continuous Parameters

We will refer to discrete input and output parameters as D_i and D_o, respectively. Accordingly, we will refer to continuous input and output parameters as $F_i(\xi)$ and $F_o(\eta)$, respectively. As shown in Chap. 2, sensitivities of any output parameters to discrete input parameters are partial derivatives:

© The Author(s) 2015
E.A. Ustinov, *Sensitivity Analysis in Remote Sensing*,
SpringerBriefs in Earth Sciences, DOI 10.1007/978-3-319-15841-9_8

$$\frac{\partial D_o}{\partial D_i}, \quad \text{and} \quad \frac{\partial F_o(\eta)}{\partial D_i} \tag{8.1}$$

And, sensitivities of any output parameters to continuous input parameters are variational derivatives:

$$\frac{\delta D_o}{\delta F_i(\xi)}, \quad \text{and} \quad \frac{\delta F_o(\eta)}{\delta F_i(\xi)} \tag{8.2}$$

Sensitivities to functions that are the continuous parameters of the models do depend on the arguments of these functions. Let ξ_1 and ξ_2 be two different types of arguments of the input functions. Then, for any type of output parameter denoted here by P_o, we have:

$$\frac{\delta P_o}{\delta F_i(\xi_2)} = \frac{\delta P_o}{\delta F_i(\xi_1)} \cdot \frac{\partial \xi_1}{\partial \xi_2} \tag{8.3}$$

The functional interdependence between ξ_1 and ξ_2 is assumed to be known, and the partial derivative $\partial \xi_1 / \partial \xi_2$ is available.

In practical applications, the continuous parameters of the models, both input and output parameters, are tabulated on suitable grids of their arguments, thus spawning the sets of corresponding discrete parameters. Sensitivities of output parameters on grid values of continuous input parameters become corresponding partial derivatives, which are expressed through the values of variational derivatives at the gridpoints $\{\xi_k\}$ as follows:

$$\frac{\partial P_o}{\partial F_i(\xi_k)} = \frac{\delta P_o}{\delta F_i(\xi)} \bigg|_{\xi=\xi_k} \cdot \Delta_k \xi \tag{8.4}$$

where $\Delta_k \xi$ is a mesh width at a gridpoint ξ_k.

In the remainder of this chapter we assume that the continuous input and output parameters are represented by their grid values. We will denote the resulting set of the grid values of input parameters by an m—vector $\mathbf{x}$. The corresponding set of the output parameters is denoted by an n—vector $\mathbf{y}$.

8.2 Error Analysis of Forward Models

Any model that is intended to be compared with observations is associated with intrinsic uncertainties. The output parameters of forward models bear uncertainties caused by errors of knowledge of the input parameters. The values of input parameters retrieved from measurements of output parameters are associated with uncertainties caused by measurement errors. In general, any quantitative result in science and technology has to be associated with its uncertainty. The science of remote sensing is, of course, not an exclusion from this rule.

8.2.1 Statistics of Multidimensional Random Variables

Consider a set of random variables $\mathbf{x}$, which obeys a multi-dimensional Gaussian distribution, specified by the average $\langle \mathbf{x} \rangle$ and a covariance matrix

$$\boldsymbol{\Sigma} = \left\langle (\mathbf{x} - \langle \mathbf{x} \rangle)(\mathbf{x} - \langle \mathbf{x} \rangle)^T \right\rangle \tag{8.5}$$

Note that by definition, the covariance matrix $\boldsymbol{\Sigma}$ is symmetric: $\boldsymbol{\Sigma}^T = \boldsymbol{\Sigma}$. This means that its inverse is also a symmetric matrix (See Appendix). This circumstance will be used in derivations below.

Up to a non-essential normalization factor, the Gaussian probability distribution function (PDF) of $\mathbf{x}$ has the form:

$$P(\mathbf{x}) \sim \exp\left(-\frac{1}{2}(\mathbf{x} - \langle \mathbf{x} \rangle)^T \boldsymbol{\Sigma}^{-1}(\mathbf{x} - \langle \mathbf{x} \rangle)\right) \tag{8.6}$$

which can be reduced to a simpler form:

$$P(\mathbf{x}) \sim \exp\left(-\frac{1}{2}\mathbf{x}^T \mathbf{W} \mathbf{x} + \mathbf{w}^T \mathbf{x}\right) \tag{8.7}$$

where

$$\mathbf{W} = \boldsymbol{\Sigma}^{-1} \tag{8.8}$$

is the inverse of the covariance matrix $\boldsymbol{\Sigma}$ of the Gaussian distribution Eq. (8.6), and thus is also symmetric, and

$$\mathbf{w} = \mathbf{W}\langle \mathbf{x} \rangle \tag{8.9}$$

is a constant vector directly associated with the average $\langle \mathbf{x} \rangle$.

In derivations below we will reduce the PDFs of random variables of question to the form of Eq. (8.6) and will use Eqs. (8.8) and (8.9) to obtain the average $\langle \mathbf{x} \rangle$ and the covariance matrix $\boldsymbol{\Sigma}$, which are used here for various applications.

8.2.2 Error Analysis of Output Parameters

When planning a remote sensing experiment, it is of obvious importance to estimate the uncertainties of the output parameters, i.e., the observables, as defined by tolerable errors of the input parameters to be retrieved. With this capability available, one can place the requirements on the errors of measurements as defined by requirements on the retrieval errors. Availability of the matrix of sensitivities—a jacobian

$$K = \frac{\partial y}{\partial x} \tag{8.10}$$

provides this capability.

We assume that the vector of measurement errors

$$\varepsilon = y - \langle y \rangle \tag{8.11}$$

obeys the Gaussian distribution with the inverse covariance matrix

$$W_y = \langle \varepsilon \varepsilon^T \rangle \tag{8.12}$$

and the average $\langle \varepsilon \rangle = 0$. Denoting $\delta x = x - \langle x \rangle$ and assuming a linear approximation, we have:

$$\varepsilon = K\delta x \tag{8.13}$$

Using the general definition of the covariance matrix Eq. (8.5) we have:

$$\Sigma_y = \langle K\delta x (K\delta x)^T \rangle \tag{8.14}$$

After a few transformations, and using Eq. (8.5), with the substitution $\delta x = x - \langle x \rangle$, we obtain the covariance matrix of the output parameters as expressed through the covariance matrix of the input parameters Σ_x and the jacobian K, which is the product of the sensitivity analysis of the model considered:

$$\Sigma_y = K\Sigma_x K^T \tag{8.15}$$

8.2.3 Error Analysis of Input Parameters

When retrieving the input parameters from measurements of the output parameters, it is of obvious importance to estimate the uncertainties of the retrieved parameters as defined by given errors of measurements. If the jacobian matrix K can be inverted, then from Eq. (8.13) we have:

$$\delta x = K^{-1}\varepsilon \tag{8.16}$$

Using the general definition, Eq. (8.5) we have:

$$\Sigma_x = \langle K^{-1}\varepsilon (K^{-1}\varepsilon)^T \rangle = K^{-1}\langle \varepsilon \varepsilon^T \rangle (K^{-1})^T = K^{-1}\Sigma_y (K^{-1})^T \tag{8.17}$$

The general case, when the jacobian K cannot be inverted directly is considered below.

8.3 Inverse Modeling: Retrievals and Error Analysis

Inverse modeling represents the essence of remote sensing. The measurements of the output parameters, observables, provides information that is used to constrain the model input parameters of interest, i.e., to obtain their estimates, along with the estimates of their uncertainties, i.e., retrieval errors.

8.3.1 General Approach to Solution of Inverse Problems in Remote Sensing

Most books on inversion methods begin with this matrix equation:

$$\mathbf{K}\mathbf{x} = \mathbf{y} \tag{8.18}$$

where the vector $\mathbf{y}$ represents the difference between the vectors of measured and modeled output parameters, and the vector $\mathbf{x}$ represents corresponding increment of the current vector of input parameters. This increment is the solution of the linearized inverse problem as represented by Eq. (8.18). Of course, the matrix of sensitivities $\mathbf{K}$ (the jacobian) as assumed to be known. The vector $\mathbf{y}$ is treated as a random variable, with the average $\langle \mathbf{y} \rangle$ corresponding to the set of measured observables and covariance matrix $\Sigma_{\mathbf{y}}$ describing the measurement errors. In the case of non-correlating measurement, $\Sigma_{\mathbf{y}}$ is a diagonal matrix

$$\Sigma_{\mathbf{y}} = \mathrm{diag}\left(s_i^2\right) \tag{8.19}$$

where the set $\{s_i\}$ represents the error measurements of the components $\{y_i\}$ of the vector $\mathbf{y}$.

The vector $\mathbf{x}$ is treated as a random variable, with the *conditional* PDF of $\mathbf{x}$, which up to a non-essential normalization factor is defined as

$$P(\mathbf{x}|\mathbf{y}) \sim \exp\left(-\frac{1}{2}\mathbf{x}^T\mathbf{W}_{\mathbf{x}}\mathbf{x} + \mathbf{w}_{\mathbf{x}}^T\mathbf{x}\right) \tag{8.20}$$

where $\mathbf{W}_{\mathbf{x}}$ and $\mathbf{w}_{\mathbf{x}}$ are derived from given parameters $\mathbf{W}_{\mathbf{y}}$ and $\mathbf{w}_{\mathbf{y}}$ of the PDF of the vector $\mathbf{y}$ and from the jacobian $\mathbf{K} = \partial\mathbf{y}/\partial\mathbf{x}$. Also, if the system of equations represented by the matrix equation Eq. (8.18) is underdefined, then some a priori information on the solutions $\mathbf{x}$ should be exploited. The cases of corresponding well-posed and ill-posed inverse problems are considered in the subsections below.

8.3.2 Well-Posed Inverse Problems and the Least Squares Method

We re-define the vector $\mathbf{y}$ as $\mathbf{y} \to \mathbf{y} + \boldsymbol{\varepsilon}$ by singling out its measurement error $\boldsymbol{\varepsilon}$. Then Eq. (8.18) takes the form:

$$\mathbf{Kx} = \mathbf{y} + \boldsymbol{\varepsilon} \tag{8.21}$$

Correspondingly, $\boldsymbol{\Sigma}_{\boldsymbol{\varepsilon}} \equiv \boldsymbol{\Sigma}_{\mathbf{y}}$, and we have:

$$P(\boldsymbol{\varepsilon}) \sim \exp\left(-\frac{1}{2}\boldsymbol{\varepsilon}^T \boldsymbol{\Sigma}_{\mathbf{y}}^{-1} \boldsymbol{\varepsilon}\right) \tag{8.22}$$

Then the conditional probability $P(\mathbf{x}|\mathbf{y})$ takes the form:

$$P(\mathbf{x}|\mathbf{y}) \equiv P(\boldsymbol{\varepsilon}) = P(\mathbf{Kx} - \mathbf{y}) \tag{8.23}$$

And, up to a non-essential normalization factor, we have:

$$P(\mathbf{x}|\mathbf{y}) \sim \exp\left(-\frac{1}{2}(\mathbf{Kx} - \mathbf{y})^T \mathbf{W}_y (\mathbf{Kx} - \mathbf{y})\right) \tag{8.24}$$

The matrix-vector expression in the argument of the exponential function in Eq. (8.24) can be transformed as follows:

$$(\mathbf{Kx} - \mathbf{y})^T \mathbf{W}_{\mathbf{y}}(\mathbf{Kx} - \mathbf{y}) = (\mathbf{Kx})^T \mathbf{W}_{\mathbf{y}}\mathbf{Kx} - \mathbf{y}^T \mathbf{W}_{\mathbf{y}}\mathbf{Kx} - (\mathbf{Kx})^T \mathbf{W}_{\mathbf{y}}\mathbf{y} + \mathbf{y}^T \mathbf{W}_{\mathbf{y}}\mathbf{y} \tag{8.25}$$

The term that is quadratic in $\mathbf{x}$ is transformed as:

$$(\mathbf{Kx})^T \mathbf{W}_{\mathbf{y}}\mathbf{Kx} = \mathbf{x}^T \mathbf{K}^T \mathbf{W}_{\mathbf{y}}\mathbf{Kx} = \mathbf{x}^T \mathbf{W}_{\mathbf{x}}\mathbf{x} \tag{8.26}$$

and we obtain:

$$\mathbf{W}_{\mathbf{x}} = \mathbf{K}^T \mathbf{W}_{\mathbf{y}}\mathbf{K} \tag{8.27}$$

which yields the covariance matrix of the solution $\mathbf{x}$ [cf. Eq. (8.8)]:

$$\boldsymbol{\Sigma}_{\mathbf{x}} = \left(\mathbf{K}^T \mathbf{W}_{\mathbf{y}}\mathbf{K}\right)^{-1} \tag{8.28}$$

The terms that are linear in $\mathbf{x}$ are transformed to an identical form, because the matrix $\mathbf{W}_{\mathbf{y}}$ is symmetric [cf. Appendix, Eq. (A.7)]:

$$\mathbf{y}^T \mathbf{W_y} \mathbf{Kx} = \left(\mathbf{K}^T \mathbf{W_y} \mathbf{y}\right)^T \mathbf{x} \tag{8.29}$$

$$(\mathbf{Kx})^T \mathbf{W_y} \mathbf{y} = \left(\mathbf{W_y} \mathbf{y}\right)^T \mathbf{Kx} = \left(\mathbf{K}^T \mathbf{W_y} \mathbf{y}\right)^T \mathbf{x} \tag{8.30}$$

Their sum is transformed as

$$\mathbf{y}^T \mathbf{W_y} \mathbf{Kx} + (\mathbf{Kx})^T \mathbf{W_y} \mathbf{y} = 2\left(\mathbf{K}^T \mathbf{W_y} \mathbf{y}\right)^T \mathbf{x} = 2\mathbf{w_x}^T \mathbf{x} \tag{8.31}$$

where the vector

$$\mathbf{w_x} = \mathbf{K}^T \mathbf{W_y} \mathbf{y} \tag{8.32}$$

yields the average of the solution $\mathbf{x}$ of Eq. (8.21) in the form [cf. Eq. (8.8), (8.9) and (8.28)]:

$$\langle \mathbf{x} \rangle = \left(\mathbf{K}^T \mathbf{W_y} \mathbf{K}\right)^{-1} \mathbf{K}^T \mathbf{W_y} \mathbf{y} \tag{8.33}$$

Equations (8.33) and (8.28) provide the least squares solution of the inverse problem Eq. (8.18).

The least squares method (LSM) works well if the non-linearity of the forward problem is not too pronounced. If the non-linearity becomes essential, then the iterations based on the use of LSM might not converge to the final solution. The method developed by Levenberg and later refined by Marquardt is more robust in such a case. A description of this method may be found elsewhere (see the References Section).

8.3.3 Ill-Posed Inverse Problems and the Statistical Regularization Method

In some sense, this method can be considered as an extension of the LSM. In a number of situations, direct application of the LSM to the inverse problem Eq. (8.2) results in a non-physical solution. For example, retrievals of profiles of atmospheric parameters specified on a fine grid might result in a solution with large variations of the grid values from one gridpoint to another.

In such cases, one remedy is in exploiting of additional, a priori information about the solution. The widely used technique is based on specifying this information in the form of its expected average $\mathbf{x}_a$ and covariance matrix $\mathbf{\Sigma}_{\mathbf{x},a}$, which are obtained from available statistics about the solution, such as long-term averages, or merely from physically reasonable assumptions about it. The off-diagonal terms of $\mathbf{\Sigma}_{\mathbf{x},a}$ describe the expected correlation between the elements of the solution $\mathbf{x}$. Then, up to a non-essential constant factor, the a priori PDF of the solution $\mathbf{x}$ has the form:

$$P_a(\mathbf{x}) \sim \exp\left(-\frac{1}{2}(\mathbf{x} - \mathbf{x}_a)^T \mathbf{W}_{\mathbf{x},a}(\mathbf{x} - \mathbf{x}_a)\right) \tag{8.34}$$

where $\mathbf{W}_{\mathbf{x},a} = \mathbf{\Sigma}_{\mathbf{x},a}^{-1}$. The conditional probability $P(\mathbf{x}|\mathbf{y})$ given by Eq. (8.24) is replaced by the product

$$P(\mathbf{x}|\mathbf{y}) \rightarrow P(\mathbf{x}|\mathbf{y})P_a(\mathbf{x}) \sim \exp\left(-\frac{1}{2}(\mathbf{K}\mathbf{x} - \mathbf{y})^T \mathbf{W}_\mathbf{y}(\mathbf{K}\mathbf{x} - \mathbf{y}) - \frac{1}{2}(\mathbf{x} - \mathbf{x}_a)^T \mathbf{W}_{\mathbf{x},a}(\mathbf{x} - \mathbf{x}_a)\right)$$
$$\tag{8.35}$$

The first term of the matrix-vector expression in the argument of the exponential function in Eq. (8.35) is identical to the matrix-vector expression in the argument of the exponential function of Eq. (8.24), and it is split into the quadratic and linear terms, as given by Eqs. (8.26), (8.29) and (8.30). The second term is transformed in an analogous way:

$$(\mathbf{x} - \mathbf{x}_a)^T \mathbf{W}_{\mathbf{x},a}(\mathbf{x} - \mathbf{x}_a) = \mathbf{x}^T \mathbf{W}_{\mathbf{x},a}\mathbf{x} - \mathbf{x}_a^T \mathbf{W}_{\mathbf{x},a}\mathbf{x} - \mathbf{x}^T \mathbf{W}_{\mathbf{x},a}\mathbf{x}_a + \mathbf{x}_a^T \mathbf{W}_{\mathbf{x},a}\mathbf{x}_a \tag{8.36}$$

Summing the terms that are quadratic in $\mathbf{x}$ given by Eq. (8.26) with those contained in Eq. (8.36), we have:

$$\mathbf{x}^T \mathbf{K}^T \mathbf{W}_\mathbf{y}\mathbf{K}\mathbf{x} + \mathbf{x}^T \mathbf{W}_{\mathbf{x},a}\mathbf{x} = \mathbf{x}^T \left(\mathbf{K}^T \mathbf{W}_\mathbf{y}\mathbf{K} + \mathbf{W}_{\mathbf{x},a}\right)\mathbf{x} \tag{8.37}$$

And we obtain:

$$\mathbf{W}_\mathbf{x} = \mathbf{K}^T \mathbf{W}_\mathbf{y}\mathbf{K} + \mathbf{W}_{\mathbf{x},a} \tag{8.38}$$

which yields the covariance matrix of the regularized solution $\mathbf{x}$:

$$\mathbf{\Sigma}_\mathbf{x} = \left(\mathbf{K}^T \mathbf{W}_\mathbf{y}\mathbf{K} + \mathbf{W}_{\mathbf{x},a}\right)^{-1} \tag{8.39}$$

The terms in Eq. (8.36), which are linear in $\mathbf{x}$, are equal to each other because the matrix $\mathbf{W}_{\mathbf{x},a}$ is symmetric:

$$\mathbf{x}_a^T \mathbf{W}_{\mathbf{x},a}\mathbf{x} = \left(\mathbf{W}_{\mathbf{x},a}^T \mathbf{x}_a\right)^T \mathbf{x} = \left(\mathbf{W}_{\mathbf{x},a}\mathbf{x}_a\right)^T \mathbf{x}, \; \mathbf{x}^T \mathbf{W}_{\mathbf{x},a}\mathbf{x}_a = \left(\mathbf{W}_{\mathbf{x},a}\mathbf{x}_a\right)^T \mathbf{x} \tag{8.40}$$

Their sum is transformed as

$$\mathbf{x}_a^T \mathbf{W}_{\mathbf{x},a}\mathbf{x} + \mathbf{x}^T \mathbf{W}_{\mathbf{x},a}\mathbf{x}_a = 2\left(\mathbf{W}_{\mathbf{x},a}\mathbf{x}_a\right)^T \mathbf{x} \tag{8.41}$$

Adding the transformed linear terms in Eqs. (8.31) and (8.41) we obtain:

$$2\left(\mathbf{K}^T\mathbf{W_y y}\right)^T\mathbf{x} + 2\left(\mathbf{W}_{x,a}\mathbf{x}_a\right)^T\mathbf{x} = 2\left(\mathbf{K}^T\mathbf{W_y y} + \mathbf{W}_{x,a}\mathbf{x}_a\right)^T\mathbf{x} = 2\mathbf{w_x}^T\mathbf{x} \qquad (8.42)$$

where

$$\mathbf{w_x} = \mathbf{K}^T\mathbf{W_y y} + \mathbf{W}_{x,a}\mathbf{x}_a \qquad (8.43)$$

yields the average of the solution $\mathbf{x}$ of Eq. (8.21) in the form [cf. Eq. (8.9)]:

$$\langle\mathbf{x}\rangle = \left(\mathbf{K}^T\mathbf{W_y K} + \mathbf{W}_{x,a}\right)^{-1}\left(\mathbf{K}^T\mathbf{W_y y} + \mathbf{W}_{x,a}\mathbf{x}_a\right) \qquad (8.44)$$

Equations (8.44) and (8.39) provide the regularized solution of the inverse problem Eq. (8.18).

References

This chapter is, essentially, a compilation of material, which can be found in a number of sources. The most prominent publication used here, is the book written by Clive Rodgers (2000). A useful complementary material can be found in Press et al. (1992).

Rodgers CD (2000) Inverse methods for atmospheric sounding: theory and practice (series on atmospheric, oceanic and planetary physics). World Scientific, Singapore

Press WS, Flannery BP, Teukolsky SA, Vetterling WT (1992) Numerical recipes in Fortran 77: the art of scientific computing. Cambridge University Press, Cambridge

Appendix
Operations with Matrices and Vectors

This Appendix contains the minimum information necessary to understand the material in this monograph involving the matrices and vectors.

A.1 Definitions

Throughout this monograph, matrices are understood as rectangular tables, composed of scalar elements consisting, in general, of m rows and n columns:

$$\mathbf{A} = \begin{pmatrix} A_{11} & A_{12} & \ldots & A_{1n} \\ A_{21} & A_{22} & \ldots & A_{2n} \\ \ldots & \ldots & \ddots & \ldots \\ A_{m1} & A_{m2} & \ldots & A_{mn} \end{pmatrix} \tag{A.1}$$

and referred to as $m \times n$ matrices. Vectors are, essentially, $n \times 1$ matrices (column $n-$ vectors, or just $n-$vectors) or $1 \times n$ matrices (row $n-$vectors):

$$\mathbf{a} = \begin{pmatrix} a_1 \\ a_2 \\ \ldots \\ a_n \end{pmatrix}, \mathbf{b} = \begin{pmatrix} b_1 & b_2 & \ldots & b_n \end{pmatrix} \tag{A.2}$$

The scalars can be considered as 1×1 matrices, or $1-$vectors.

The transpose $\mathbf{A}^T$ of matrix $\mathbf{A}$, Eq. (A.1), is an $n \times m$ matrix with elements

$$\left(\mathbf{A}^T \right)_{ij} = A_{ji} \ (i = 1, \ldots m, j = 1, \ldots n) \tag{A.3}$$

© The Author(s) 2015
E.A. Ustinov, *Sensitivity Analysis in Remote Sensing*,
SpringerBriefs in Earth Sciences, DOI 10.1007/978-3-319-15841-9

In particular, vectors **a** and **b** in Eq. (A.2) are mutually transpose to each other if $a_i = b_i$ $(i = 1, \ldots n)$:

$$\mathbf{b} = \mathbf{a}^T, \ \mathbf{a} = \mathbf{b}^T \tag{A.4}$$

The square matrix **A** is called a symmetric matrix if $\mathbf{A}^T = \mathbf{A}$.

Block matrices are defined as matrices, elements of which are matrices themselves (or vectors and scalars, in particular). These elements should satisfy two requirements: (1) matrix elements of each row of the block matrix should have equal numbers of rows; (2) matrix elements of each column of the block matrix should have equal numbers of columns. In particular, an $m \times n$ matrix can be considered both as a column $m-$vector consisting of m row $n-$vectors, and as a row $n-$vector consisting of n column $m-$vectors.

A.2 Algebra of Matrices and Vectors

Matrices with matching numbers of rows and columns are added to, and subtracted from each other just element by element:

$$(\mathbf{A} \pm \mathbf{B})_{ij} = (\mathbf{A})_{ij} \pm (\mathbf{B})_{ij} \tag{A.5}$$

The elements of the product of a matrix **A** with m columns and a matrix **B** with m rows are defined as a sum of m terms:

$$(\mathbf{AB})_{ij} = \sum_{k=1}^{m} (\mathbf{A})_{ik} (\mathbf{B})_{kj} \tag{A.6}$$

The resulting matrix **AB** has the number of rows corresponding to that of the matrix **A** and number of columns corresponding to that of the matrix **B**. The number of rows of the matrix **A** and number of columns of the matrix **B** can be arbitrary. In general, the matrix product is not commutative, and $\mathbf{AB} \neq \mathbf{BA}$.

By direct substitution it can be seen that

$$(\mathbf{AB})^T = \mathbf{B}^T \mathbf{A}^T, \ (\mathbf{ABC})^T = \mathbf{C}^T \mathbf{B}^T \mathbf{A}^T, \text{ etc.} \tag{A.7}$$

Important particular cases are presented below.

The product **ab** of the row $n-$vector **a** and the column $n-$vector **b** is a scalar:

$$\mathbf{ab} = \sum_{k=1}^{n} (\mathbf{a})_k (\mathbf{b})_k \tag{A.8}$$

The product **ba** of these vectors is an $n \times n$ matrix with elements:

$$(\mathbf{ba})_{ij} = (\mathbf{b})_k (\mathbf{a})_k, \ (i, j = 1, \ldots n) \tag{A.9}$$

The product $a\mathbf{b}$ of the scalar a and the row n−vector **b** is a row n−vector with elements:

$$(a\mathbf{b})_j = a(\mathbf{b})_j \tag{A.10}$$

The product $\mathbf{a}b$ of the column n−vector **a** and the scalar b is a column n−vector with elements:

$$(\mathbf{a}b)_j = (\mathbf{a})_j b \tag{A.11}$$

In the remainder of this Appendix, speaking of vectors we mean column vectors.

The scalar, or inner, product $(\mathbf{a}, \mathbf{b}) \equiv \mathbf{a} \cdot \mathbf{b}$ of two n−vectors **a** and **b** is defined as a scalar:

$$c = \mathbf{a} \cdot \mathbf{b} = \mathbf{a}^T \mathbf{b} = \sum_{k=1}^{n} (\mathbf{a})_k (\mathbf{b})_k \tag{A.12}$$

The tensor, or outer, product $\mathbf{a} \otimes \mathbf{b}$ of an m−vector **a** and n−vector **b** is defined as an $m \times n$ matrix

$$\mathbf{C} = \mathbf{a} \otimes \mathbf{b} = \mathbf{a}\mathbf{b}^T \tag{A.13}$$

with elements

$$(\mathbf{C})_{ij} = (\mathbf{a})_i (\mathbf{b})_j \tag{A.14}$$

The identity matrix **I** is a square matrix with diagonal elements $(\mathbf{I})_{ii} \equiv 1$ and non-diagonal elements $(\mathbf{I})_{i \neq j} \equiv 0$. It commutes with any square matrix **A** of matching size:

$$\mathbf{AI} = \mathbf{IA} = \mathbf{A} \tag{A.15}$$

The inverse matrix $\mathbf{A}^{-1}$ of the square matrix **A** is defined as a square matrix satisfying the equality

$$\mathbf{A}\mathbf{A}^{-1} = \mathbf{I} \tag{A.16}$$

By definition $\mathbf{A}^{-1}$ commutes with **A**.

The transpose and inversion operations are commutative:

$$\left(\mathbf{A}^T\right)^{-1} = \left(\mathbf{A}^{-1}\right)^T \tag{A.17}$$

The simple proof of this consists in reduction of Eq. (A.17) to an identity $\mathbf{I} = \mathbf{I}$ by identical transformations: multiplication of both sides of Eq. (A.17) by $\mathbf{A}^T$, application of Eq. (A.7) to $\mathbf{A}$ and $\mathbf{A}^{-1}$, and application of Eq. (A.16) to pairs $\mathbf{A}, \mathbf{A}^{-1}$ and $\mathbf{A}^T, \left(\mathbf{A}^T\right)^{-1}$ in both sides of the resulting equality.

A.3 Differential Operations

Derivatives of matrices and vectors with respect to a single scalar argument are carried out element by element. In the general case of an $m \times n$ matrix, Eq. (A.1) the derivative with respect to an argument x is an $m \times n$ matrix with elements:

$$\left(\frac{d\mathbf{A}}{dx}\right)_{ij} = \frac{d(\mathbf{A})_{ij}}{dx} \tag{A.18}$$

Derivatives with respect to spatial coordinates $\mathbf{r}$ in 2D and 3D space, which are combined, correspondingly, in a 2– or 3– vector

$$\mathbf{r} = \begin{pmatrix} x \\ y \end{pmatrix}, \text{ or } \mathbf{r} = \begin{pmatrix} x \\ y \\ z \end{pmatrix} \tag{A.19}$$

are represented by the nabla operator ∇, which is, correspondingly, a 2– or 3– vector

$$\nabla = \frac{\partial}{\partial \mathbf{r}} = \begin{pmatrix} \partial/\partial x \\ \partial/\partial y \end{pmatrix}, \text{ or } \nabla = \frac{\partial}{\partial \mathbf{r}} = \begin{pmatrix} \partial/\partial x \\ \partial/\partial y \\ \partial/\partial z \end{pmatrix} \tag{A.20}$$

The nabla operator can be applied both to scalars and vectors following the multiplication rules presented above. If $a(\mathbf{r})$ is a scalar function, then

$$\nabla a = \frac{\partial a}{\partial \mathbf{r}} = \begin{pmatrix} \partial a/\partial x \\ \partial a/\partial y \end{pmatrix}, \text{ or } \nabla a = \frac{\partial a}{\partial \mathbf{r}} = \begin{pmatrix} \partial a/\partial x \\ \partial a/\partial y \\ \partial a/\partial z \end{pmatrix} \tag{A.21}$$

is, correspondingly, a 2– or 3– vector. The result is a gradient of a: $\nabla a = \text{grad } a$. If $\mathbf{a(r)}$ is a 2– or 3– vector function, then

$$\nabla^T \mathbf{a} = \left(\frac{\partial}{\partial \mathbf{r}}\right) \cdot \mathbf{a} = \frac{\partial a_x}{\partial x} + \frac{\partial a_y}{\partial y}, \text{ or } \nabla^T \mathbf{a} = \left(\frac{\partial}{\partial \mathbf{r}}\right) \cdot \mathbf{a} = \frac{\partial a_x}{\partial x} + \frac{\partial a_y}{\partial y} + \frac{\partial a_z}{\partial z} \quad (A.22)$$

The result is a divergence of $\mathbf{a}$: $\nabla^T \mathbf{a} = \text{div } \mathbf{a}$.

Derivatives of products of matrices and vectors are, in suitable notations, the straight forward generalizations of the case of scalar functions of a single scalar argument:

$$(uv)' = u'v + uv' \quad (A.23)$$

For the general case of the product of a matrix $\mathbf{A}$ with m columns and a matrix $\mathbf{B}$ with m rows, by direct substitution it can be seen that the elements of the derivative with respect to a single scalar argument have the form [cf. Eq. (A.6)]:

$$(\mathbf{AB})' = \mathbf{A}'\mathbf{B} + \mathbf{AB}' \quad (A.24)$$

A.4 Integral Operations

Integration of matrices and vectors with respect to any scalar argument are carried out element by element. In the general case of an $m \times n$ matrix $\mathbf{A}$, Eq. (A.1) the integral over an argument x is an $m \times n$ matrix with elements:

$$\left(\int \mathbf{A} \, dx\right)_{ij} = \int (\mathbf{A})_{ij} \, dx \quad (A.25)$$

Integration over a multidimensional domain D_x of arguments x is reduced to a corresponding multiple integration:

$$\left(\int_{D_x} \mathbf{A} \, dx\right)_{ij} = \int_{D(x)} (\mathbf{A})_{ij} \, dx \quad (A.26)$$

By analogy with the inner product (a, b) of two scalar functions $a(x)$ and $b(x)$

$$(a, b) = \int_{D_x} a(x)b(x) \, dx \quad (A.27)$$

the inner product of two vector functions $\mathbf{a}(x)$ and $\mathbf{b}(x)$ has the form:

$$(\mathbf{a}, \mathbf{b}) = \int_{D_x} \mathbf{a}^T(x)\mathbf{b}(x) \, dx \qquad (A.28)$$

Similarly, the inner product of two matrices $\mathbf{A}$ and $\mathbf{B}$ with a matching number of rows has the form:

$$(\mathbf{A}, \mathbf{B}) = \int_{D_x} \mathbf{A}^T(x)\mathbf{B}(x) \, dx \qquad (A.29)$$

In particular, the inner product of an $n \times m$ matrix $\mathbf{A}$ and an $n-$ vector $\mathbf{b}$ has the form:

$$(\mathbf{A}, \mathbf{b}) = \int_{D_x} \mathbf{A}^T(x)\mathbf{b}(x) \, dx \qquad (A.30)$$

Index

A

Adjoint approach, 14, 16, 19, 20, 23, 30, 35, 42, 56, 65, 74

Adjoint operator, 14–16, 30, 36, 45, 56, 57, 65, 66, 74, 79, 81, 100, 108

Adjoint problem, 14, 16, 30, 31, 36, 42, 44, 45, 56, 66, 68, 74–76, 78–80, 82–88, 90, 91, 93, 94, 97, 105, 106, 109, 110

Adjoint propagator, 75, 76

Adjoint solution, 31, 46, 58, 75, 92, 94, 98, 106, 110

B

Baseline forward problem, 15, 16, 28, 29, 31, 34, 49, 51, 53, 55, 56, 58, 61, 67

Baseline solution, 12, 53, 63, 71

Bi-harmonic equation, 87

Boundary conditions, 4, 15, 42, 43, 45, 53, 54, 58, 78, 83, 87, 90

Boundary of domain, 43, 45, 82, 102

C

Continuous parameters, 4, 5, 7, 15, 38, 40, 47, 53, 54, 58, 59, 111, 112

Covariance matrix, 113–118

D

Delta function, 63

Differential calculus, 28, 32–34, 37

Differential equation, 4, 15, 18, 19, 21, 27, 28, 31, 35, 38, 45, 61, 68, 69, 75, 78, 80, 89

Dirichlet boundary condition, 90

Discrete parameters, 5, 6, 22, 38, 46, 53, 58, 111, 112

Domain of arguments, 5

E

Error analysis, 112–115

F

Final-value condition, 80

Finite-difference approach, 11, 13

Forward model, 2–4, 12, 16, 112

Forward operator, 58

Forward problem, 4, 5, 10, 11, 13–17, 19–21, 27–31, 33–36, 38, 40, 42, 46, 49, 50, 53–55, 65, 78, 80, 83, 85, 87, 90, 91, 94, 96, 99, 100, 106, 110, 117

Forward solution, 4–6, 11–13, 17–19, 28, 29, 56, 76, 80, 82–84, 86, 88, 96, 100

H

Heat equation, 90, 92

I

Identity matrix, 63–65, 69, 83, 105

Ill-posed inverse problems, 115, 117

Indefinite conditions, 78, 83, 85, 86, 88, 89, 91, 93, 97, 98, 100, 107, 109

Initial-value condition, 80

Intensity of radiation, 20

Inverse modeling, 115

J

Jacobian, 15, 113–115

L

Lagrange identity, 14, 30, 43–45, 56, 65, 66, 68, 74, 77–81, 100, 108

Lambda function, 43, 45

Laplacian, 99, 106

© The Author(s) 2015

E.A. Ustinov, *Sensitivity Analysis in Remote Sensing*, SpringerBriefs in Earth Sciences, DOI 10.1007/978-3-319-15841-9